This Book Belongs to:

....

*Thanks to Mum and Dad for planting the 'seed' in me, Dawn Sanders,
Anna Lewington, Kew and Chelsea Physic Garden for helping
the seedling to grow and Karen, Serena and friends for watching it flower.
Nature, if you're listening, thanks too!
– Michael Holland*

*To my precious flowers, Nina and Leo.
– Philip Giordano*

I Ate Sunshine for Breakfast is © Flying Eye Books 2020

This is a first edition published in 2020 by Flying Eye Books,
an imprint of Nobrow Ltd. 27 Westgate Street, London E8 3RL.

Text © Michael Holland FLS 2020
Illustrations © Philip Giordano 2020

Book Design by Bia Melo

Michael Holland has asserted his right under the Copyright,
Designs and Patents Act, 1988, to be identified as the Author of this Work.
Philip Giordano has asserted his right under the Copyright,
Designs and Patents Act, 1988, to be identified as the Illustrator of this Work.

Illustrations published by arrangement with Debbie Bibo Agency

Every attempt has been made to ensure any statements written as fact have been checked
to the best of our abilities. However, we are still human, thankfully, and occasionally
little mistakes may crop up. Should you spot any errors, please email info@nobrow.net.

3 5 7 9 10 8 6 4 2

Published in the US by Nobrow (US) Inc.

Printed in Latvia on FSC® certified paper.

FSC — MIX — Paper from responsible sources — FSC® C002795

ISBN: 978-1-911171-18-8

www.flyingeyebooks.com

Safety note: We advise that all activities undertaken in this book are completed with adult supervision. 95% of plant species are poisonous, and whilst this book mentions some of these, it is not exhaustive. If you have allergies, including asthma, avoid handling any plants mentioned in this book, or wear gloves and a face mask. Always make sure to wash your hands thoroughly after coming in to contact with any plants. Speak to a medically trained herbologist if you plan to take plants for medicinal purposes.

Michael Holland FLS & Philip Giordano

I ATE SUNSHINE FOR BREAKFAST

A Celebration of Plants Around the World

Flying Eye Books
London | New York

CONTENTS

Part One:
ALL ABOUT PLANTS

10 **Why do plants matter?**
12 **What is a plant?**
14 **Plant parts**
16 **Leaves: a food factory**
18 DIY: Make your own plant maze
20 **Flower power**
22 **Parts of a flower**
24 **Pollination**
26 DIY: Wild weed bottle garden
28 **The birth of a plant**

30 **Seeds on the move**
32 DIY: Play conkers
34 **Living fossils**

Part Two:
WORLD OF PLANTS

38 **The plant kingdom**
40 **Happy families**
42 DIY: Cornflour slime
44 **Evolution**
46 **Adaptation**
48 **Life in extremes: Hot and dry**
50 **Life in extremes: In the jungle**

52 **Watery world**
54 **Plant tricks**
56 DIY: Evergreen freeze
58 **Poison**
60 **Food chains and food webs**
62 **It's a trap!**

Part Three:
FROM BREAKFAST
UNTIL BEDTIME

66 I ate sunshine for
    breakfast – and so
    did you!
68 Plant juice
70 Time to brush
    your teeth
72 Clean up!
74 DIY: Shelf Life Project
76 Let's get dressed
78 Colourful world
80 Sweet and sour
    smells
82 DIY: Leaf printing

84 Around the house
86 Pencil and paper
88 Strike up the band
90 DIY: Grass squeaker
92 Sporting life
94 DIY: Bean bag boules

Part Four:
THE POWER
OF PLANTS

98 Smart plant
    technology
100 DIY: Potato
    power plant
102 Hunting and fighting
104 DIY: Invisible ink
106 Green healing

108 Speaking in plant
110 Get going!
112 Pollution
114 Taking care
    of the planet
116 DIY: Local living
    landmarks
118 The future is green
120 Plant awards
122 Glossary
124 Index

Part One

ALL ABOUT PLANTS

Plants are essential to your world. Without them, no other living thing would be able to survive. This book will help you to become more acquainted with your leafy neighbours, from how they grow, to fossilisation and everything in-between. With over 400,000 species on our planet to discover, let's find out all about these incredible organisms!

WHY DO PLANTS MATTER?

Every single day, and in every single way, we use plants. From the food we eat and the cars we drive to the medicines we take and the clothes we wear, we would not be able to live without plants.

At this very moment you are holding a book made from plants. And not just one plant – many plants! The pages and covers of this book are made from a mixture of two types of tree – birch *(Betula)* and pine *(Pinus)*. The words you are reading are printed with ink made from soybeans *(Glycine max)* and linseed/flax oil *(Linum usitatissimum)*.

Plants touch almost every aspect of our lives. There are about 428,000 **species** of plants growing on our amazing planet and 34,000 of these have recorded benefits. Many people spend years researching plants and their uses. This scientific study is called **ethnobotany**.

In the pages of this book you'll learn exactly what plants are, how they work, and all the ingenious ways you use plants every day. Along the way, there are plenty of cool plant experiments to try for yourself.

Soybean

10

Pine

Silver birch

Flax

A NOTE ON PLANT NAMES:

This book gives both the common and scientific names of plants. The scientific name (shown in italics below) helps us know exactly which plant we're talking about. Its 'genus name' comes first (and is like a family name), followed by its 'species name' (which is like a first name), like this:

Silver birch (*Betula pendula*)

11

WHAT IS A PLANT?

A plant is a living thing that usually grows in a permanent place. The plant kingdom is so varied that it includes microscopic algae and pretty flowers, right up to huge trees that can live for thousands of years. It survives by taking in water (H_2O) and nutrients through its roots and 'breathing' in carbon dioxide (CO_2) through its leaves. Using sunlight energy, a plant can also make sugary food in a process known as **photosynthesis**. Plants also need oxygen to survive. They use it to break down their food, which helps them grow (see pages 16-17). This is a process called respiration.

12

DID YOU KNOW?

It takes just 500 seconds for the Sun's energy to reach Earth. To reach the Sun's surface from its core, it takes a staggering 20,000 years!

PLANT PARTS

Each part of a plant has a specific job to do. Let's take a closer look at this poppy and its defining features to give us more insight into how plants work.

Leaves

These thin, flat organs are where tiny **chloroplasts** can be found. These make the leaves look green. Leaves also help the plant to make its own food. Learn more about this on pages 16-17.

Stem

Flexible and strong, a stem, stalk or trunk helps the plant to stand up from the ground. It also protects the important transportation tubes that carry sugary food down from the plant's leaves and water up from its roots.

Roots

Roots seek out and suck up water and nutrients. They also anchor the plant into the soil.

Flowers

Most plants produce flowers at some point in their life, so that they can make seeds and therefore more plants. Learn more about flowers on pages 20-23.

Poppy

15

LEAVES: A FOOD FACTORY

All life on Earth is part of a worldwide web of energy flow – starting with our closest star, the Sun. Leaves have the amazing ability to capture sunlight energy and turn it into food in a process known as photosynthesis. In this process, leaves turn sunlight energy, water, minerals and carbon dioxide into food.

Eating sunshine

The process starts with a green substance called **chlorophyll**, which is kept in tiny structures inside a plant's leaves called 'chloroplasts'. Chlorophyll absorbs energy from sunlight and turns it into carbohydrates, which are a type of sugar. Along with minerals from the soil, this is what plants need to grow.

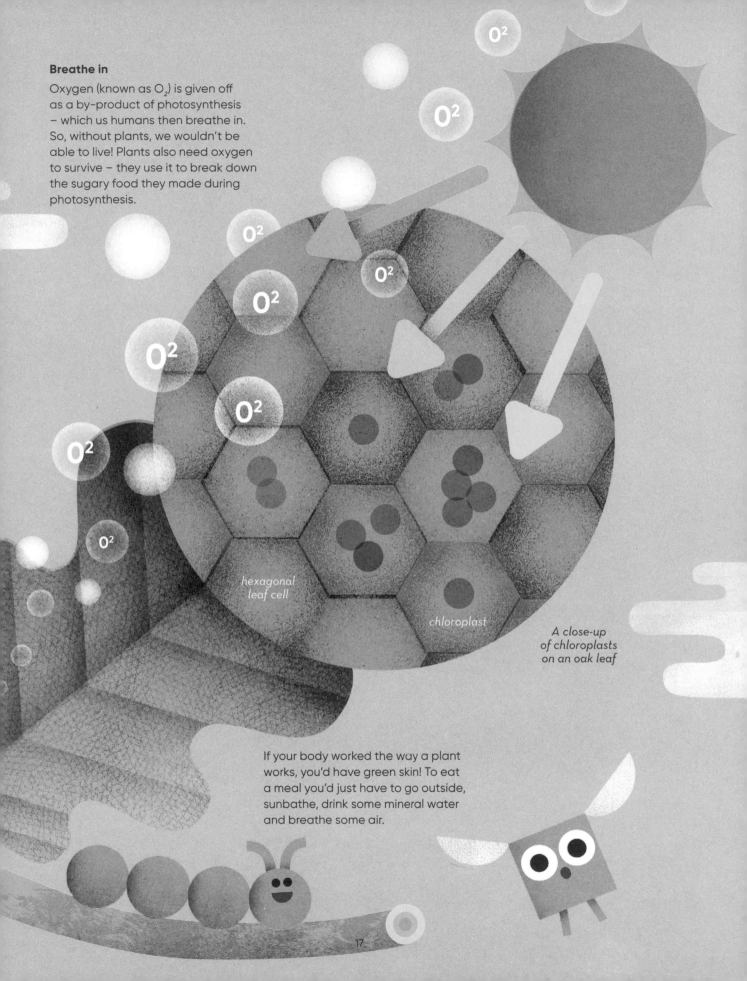

Breathe in

Oxygen (known as O_2) is given off as a by-product of photosynthesis – which us humans then breathe in. So, without plants, we wouldn't be able to live! Plants also need oxygen to survive – they use it to break down the sugary food they made during photosynthesis.

hexagonal leaf cell

chloroplast

A close-up of chloroplasts on an oak leaf

If your body worked the way a plant works, you'd have green skin! To eat a meal you'd just have to go outside, sunbathe, drink some mineral water and breathe some air.

17

DIY: MAKE YOUR OWN PLANT MAZE

This activity is a good demonstration of how plants actively seek light to survive. Once the seed has been planted, over the next few weeks a shoot should make its way out of the maze and towards the light. This process is known as **phototropism**. A good place to try this experiment is on a window sill – somewhere indoors with lots of sunshine.

Safety note: ask an adult to help you make the holes.

You Will Need:

☐ A large shoebox with a lid

☐ Stiff cardboard

☐ Scissors

☐ Bean seeds (climbing or French beans are ideal)

☐ Seed compost in a 9cm diameter plant pot

☐ A small plant saucer (or a jar lid will do)

☐ Strong sticky tape

How to Make Your Maze:

1. Cut a hole in one end of the shoebox.

3. Prop a few different pieces of card inside the shoebox. Use a bit of tape to attach them so the card forms a maze.

2. Get some card that is wider than your shoebox. Cut some simple shapes in it.

5. Put the shoebox lid back on and carefully put the whole box on a windowsill. In a few weeks you'll hopefully spot a green plant coming out of the top.

4. Plant your seeds in the compost and put the plant at the bottom of the shoebox.

FLOWER POWER

Flowers are the parts of a plant that enable it to reproduce. The vibrant petals on the outside act like advertisements to insects, and sometimes birds, for the main attraction in the middle of the flower – sweet nectar. Enticed by the sights and smells these creatures, known as **pollinators**, hop from one flower to another, taking pollen with them, and helping the flowers to produce seeds. See pages 24–25 to learn more about how this works.

Bindweed

Daisy

Tulip

Foxglove

Primrose

Artichoke

Lily of
the valley

PARTS OF A FLOWER

Most of us are familiar with petals and stems, but there is so much more to know about the parts of a flower, and what job each part has to do. Let's take a closer look at this apple blossom.

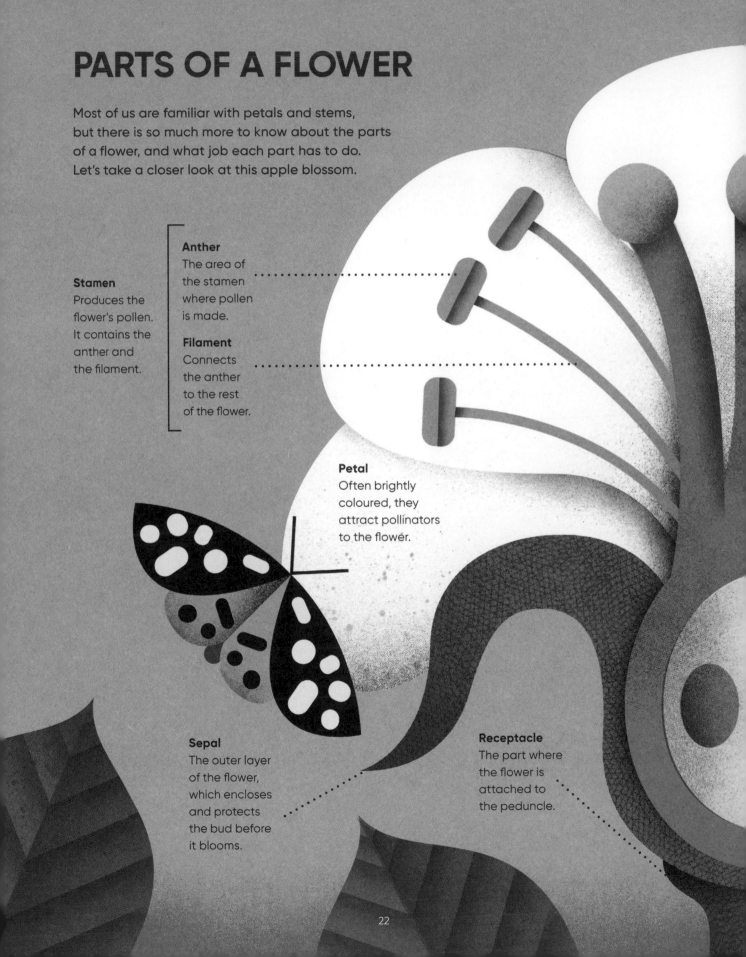

Anther
The area of the stamen where pollen is made.

Stamen
Produces the flower's pollen. It contains the anther and the filament.

Filament
Connects the anther to the rest of the flower.

Petal
Often brightly coloured, they attract pollinators to the flower.

Sepal
The outer layer of the flower, which encloses and protects the bud before it blooms.

Receptacle
The part where the flower is attached to the peduncle.

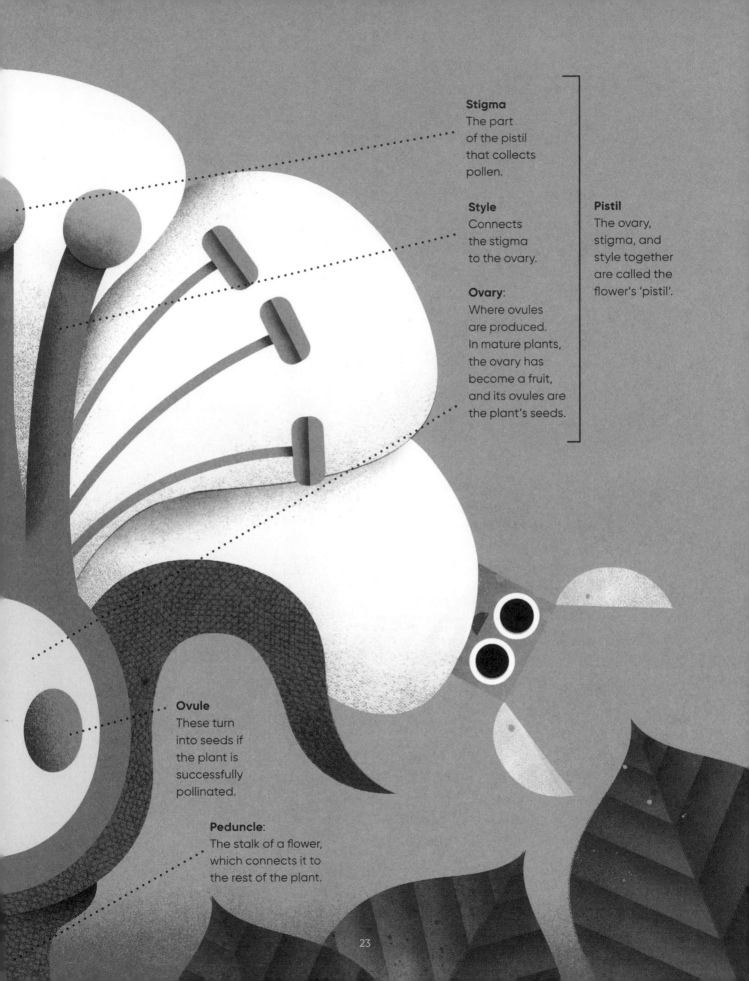

Stigma
The part of the pistil that collects pollen.

Style
Connects the stigma to the ovary.

Ovary:
Where ovules are produced. In mature plants, the ovary has become a fruit, and its ovules are the plant's seeds.

Pistil
The ovary, stigma, and style together are called the flower's 'pistil'.

Ovule
These turn into seeds if the plant is successfully pollinated.

Peduncle:
The stalk of a flower, which connects it to the rest of the plant.

23

POLLINATION

Pollination is the process in which a plant's pollen is moved from one flower to another so that seeds can form and new plants can grow. Sometimes when pollen from one species is transferred to a sister plant, a new species is created! This is known as cross-pollination. Scientists use this method to create plants that have frillier petals, richer scents, brighter colours or a whole host of other features. This is similar to the way that humans have selectively bred dogs to have different personalities or looks.

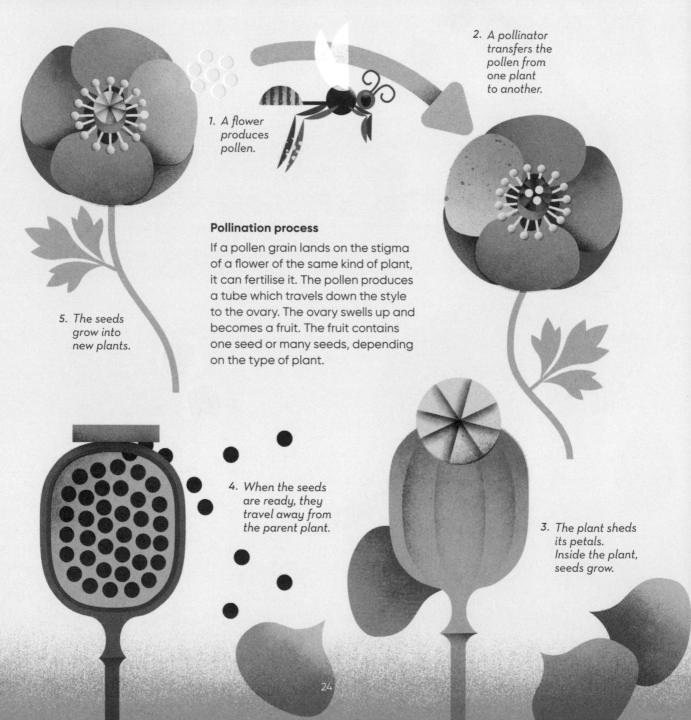

2. A pollinator transfers the pollen from one plant to another.

1. A flower produces pollen.

Pollination process

If a pollen grain lands on the stigma of a flower of the same kind of plant, it can fertilise it. The pollen produces a tube which travels down the style to the ovary. The ovary swells up and becomes a fruit. The fruit contains one seed or many seeds, depending on the type of plant.

5. The seeds grow into new plants.

4. When the seeds are ready, they travel away from the parent plant.

3. The plant sheds its petals. Inside the plant, seeds grow.

Bat feeding at a baobab flower

Pollinators

There are lots of ways a plant can be pollinated. Perhaps the most well-known is by animals, including insects, bats and birds, which brush against plants while getting to the sweet nectar inside. In doing so, they rub pollen from one flower on to another. Have you ever noticed busy buzzing bees hopping from flower to flower in the summertime? Bees are incredibly important in pollination and are 'employed' by human fruit growers to pollinate crops such as apples, tomatoes, almonds and cucumbers.

Bees pollinating the flowers of a tomato plant

Hummingbird reaching into a Heliconia flower

DIY: WILD WEED BOTTLE GARDEN

A weed is a wild plant growing in the 'wrong' place, which means it's in competition with plants growing in the 'right' place. This means any plant can be a weed. Their seeds can lay dormant in the soil for years, waiting for the ideal conditions in which to suddenly appear – as this experiment will hopefully demonstrate.

You Will Need:

- ☐ 500ml natural soil (not shop-bought)

- ☐ A 2 litre, (or larger) clear plastic bottle (soak in warm water first to remove the label)

- ☐ Funnel (optional), or cone of paper

How to Make Your Bottle Garden:

1. Making sure there aren't any animals in the soil you have, quarter-fill your bottle with soil. You may want to use a funnel to help. If dry, moisten the soil with a little water, but if it is already damp, you won't need to water it.

2. Place the bottle garden in a cool, light place, pop the lid back on and... wait!

3. You may find at first that the soil seems to have few or no seeds. Be patient. After they germinate (see page 28) you might be surprised at how many seeds there must have been in the 'empty' soil.

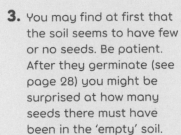

4. Try collecting soil from three or more different gardens and comparing what grows from the different samples. Can you identify the plants in them? How different are they?

27

THE BIRTH OF A PLANT

If a healthy seed is planted in a good place at the right time, it will grow into a plant. This is called germination. The seed contains all the food it needs to germinate.

Seed coat

A seed is planted in the soil.

A root appears and grows down into the soil.

The root develops root hairs and begins to push the seed upwards.

The seed splits out of its seed coat and develops a shoot.

The Germinator

All seeds need the same things to germinate – warmth, oxygen, moistness and darkness. That's why we bury them underground. If they receive enough water and warmth, they will start to grow.

Seeds will adapt to their environment. For example, parts of Australia and South Africa often have forest fires, so some plants there have seeds that need either the heat of a fire to melt their fruits and release them, or smoke to signal the seed to start growing. It's sort of like an alarm clock for germination! Hawthorn (Crataegus) seeds have such a tough outer shell, they need to be in the ground for two winters before a root can break through and the plant can begin to grow.

World's biggest seed

The coco de mer (*Lodoicea*) is the largest seed in the world and can weigh up to 22kg (48.5lb). That's about the weight of three bowling balls. It can take more than five years to ripen and grows on just two islands in the Seychelles (a country made up of many islands in the Indian Ocean).

The shoot becomes a stem and develops leaves. It has become a plant.

DID YOU KNOW?

Because it is big and brown, some people say the coco de mer seed looks like a gorilla's bottom!

SEEDS ON THE MOVE

If a seed fell right at its parent plant's roots and tried to grow, it would be sharing food, light, water and space with its parent and siblings. This means it needs to move away from home. Plants have evolved different ways of **dispersing** their seeds, or transporting them to new places where it will be easier for them to grow.

Dandelion

Dandelion seed

Birch tree fruit

Birch tree seed

Sycamore seed

By air

The signature fluffy white parachutes of the dandelion (Taraxacum officinale) are connected to its seeds. When the seeds are ready to go, the fine white hairs help the wind pick up the parachutes and carry them away. This could be just metres away, or many kilometres. There are lots of types of flying seeds. Sycamore (Acer pseudoplatanus), ash (Fraxinus) and maple tree (Acer) seeds are shaped like helicopter blades and can twirl through the air. Thin birch tree seeds can glide on the wind.

By water

Coconuts are actually giant seeds, and they travel by sea! Their hairy outside 'lifevest' helps them float and keeps saltwater out as they travel to new lands, sometimes across hundreds of kilometres. They can be seen growing on many **tropical** beaches.

Coconut

Some seeds will not survive the journey and may be broken, burnt or eaten along the way. This is why plants make lots of them, because not all of them will grow into plants and spread more seeds. But how do seeds travel? Although they come from plants, which plainly cannot walk, seeds travel in somewhat similar ways to us.

By animal

Animals help to scatter seeds just by moving about. Many plants have delicious fruits with seeds inside. When an animal eats the seeds, they stay in its body until they are pooped out later (egested). The seeds of fruits like cherry or tomato have very tough coats, and they need the digestive juices of animals to break them down so that they can germinate when they land.

Burdock fruits (*Arctium*) use a different method. They are covered in tiny hooks which attach themselves to fur for long-distance travelling. The designer of Velcro looked at burdock fruits for inspiration. This is an example of **biomimicry** (see p.73).

Squirting cucumber

Burdock fruit

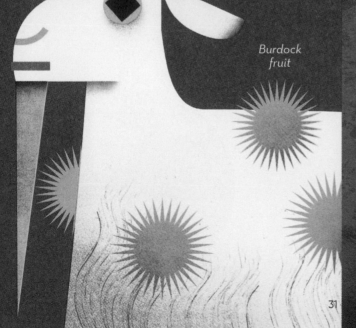

By explosion

Some seeds can move all by themselves! Well, sort of. Plants with pods or capsules dry up and can suddenly split open, catapulting their seeds into the air. This is called an **explosive** mechanism. This movement can be seen in the twisting pods of some members of the bean family, or witch hazel. The squirting cucumber (*Ecballium elaterium*) shoots its seeds as far away as six metres.

29

DIY: PLAY CONKERS

This game is perfect to play in temperate regions in the autumn, when lots of fallen horse chestnut seeds (*Aesculus hippocastanum*) can be found. Just make sure to be careful when you are swinging the conker around!

You Will Need:

- ☐ A long piece of string, or a shoelace (about 30 cm)
- ☐ Some seeds from a horse chestnut tree (conkers)
- ☐ An awl (for piercing holes)
- ☐ Someone to play with!

How to Make Your Conkers:

 1. Collect and clean your conkers.

2. Using an awl, pierce a hole all the way through the conker, from top to bottom. Ask an adult to do this for you.

3. Thread your string or shoelace through the hole and tie a knot securely at the base.

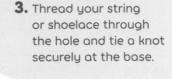

4. Repeat steps 1-3, so your opponent also has a piece to play with.

5. Taking turns, swing the conker on the string and try to hit your opponent's conker until it is destroyed. The one that remains intact, wins!

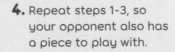

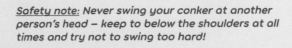

Safety note: _Never swing your conker at another person's head – keep to below the shoulders at all times and try not to swing too hard!_

33

LIVING FOSSILS

There are some species of plants living today that have been around for so long that the dinosaurs lived among them. Some of these survivors are quite incredible.

Ancient plants still pop up in some surprising places. For example, stunt explosions in movies use powdered clubmosses for a fiery finish. Creamy ice-cream is made thicker using seaweeds, while dyes can be made from lichens. Many plants including Ginkgo biloba and horsetails are used as medicines. Here are some other examples of plants that have been around from very early on in the evolutionary journey of life.

Wollemi pine

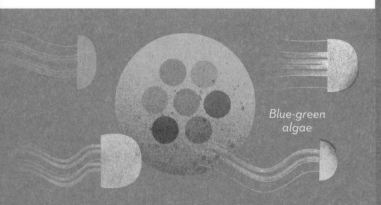

Blue-green algae

Ancient algae

During the Silurian era, which began around 433 million years ago, simple algae lived in warm, shallow oceans. The single-celled blue-green algae breathed out oxygen. This built up in the atmosphere and allowed larger creatures to evolve. This species is still pumping out oxygen to this day and is one of the most useful organisms in the world. Each one is only about a tenth of a millimetre wide, but can multiply very quickly and create rafts that spread for thousands of kilometres.

34

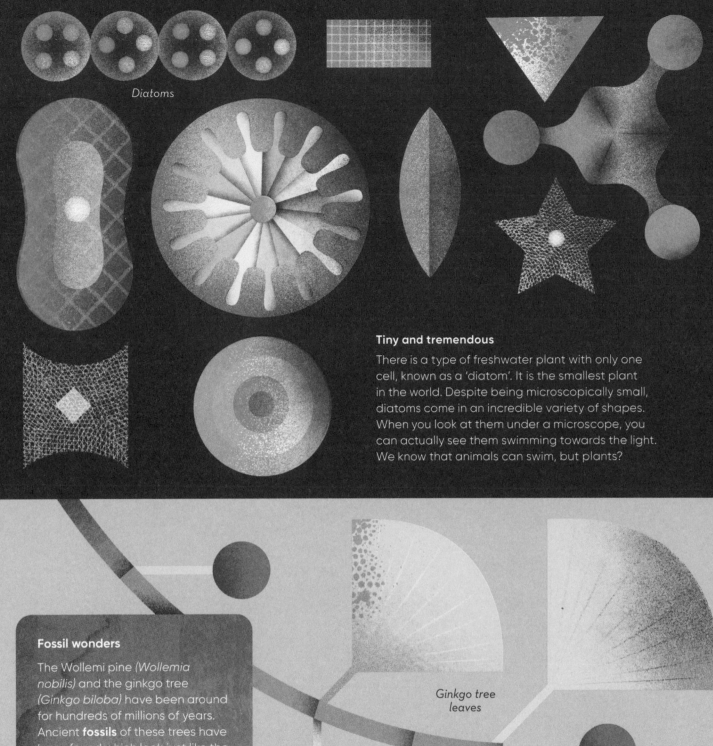

Diatoms

Tiny and tremendous

There is a type of freshwater plant with only one cell, known as a 'diatom'. It is the smallest plant in the world. Despite being microscopically small, diatoms come in an incredible variety of shapes. When you look at them under a microscope, you can actually see them swimming towards the light. We know that animals can swim, but plants?

Fossil wonders

The Wollemi pine *(Wollemia nobilis)* and the ginkgo tree *(Ginkgo biloba)* have been around for hundreds of millions of years. Ancient **fossils** of these trees have been found which look just like the ones living today!

Ginkgo tree leaves

35

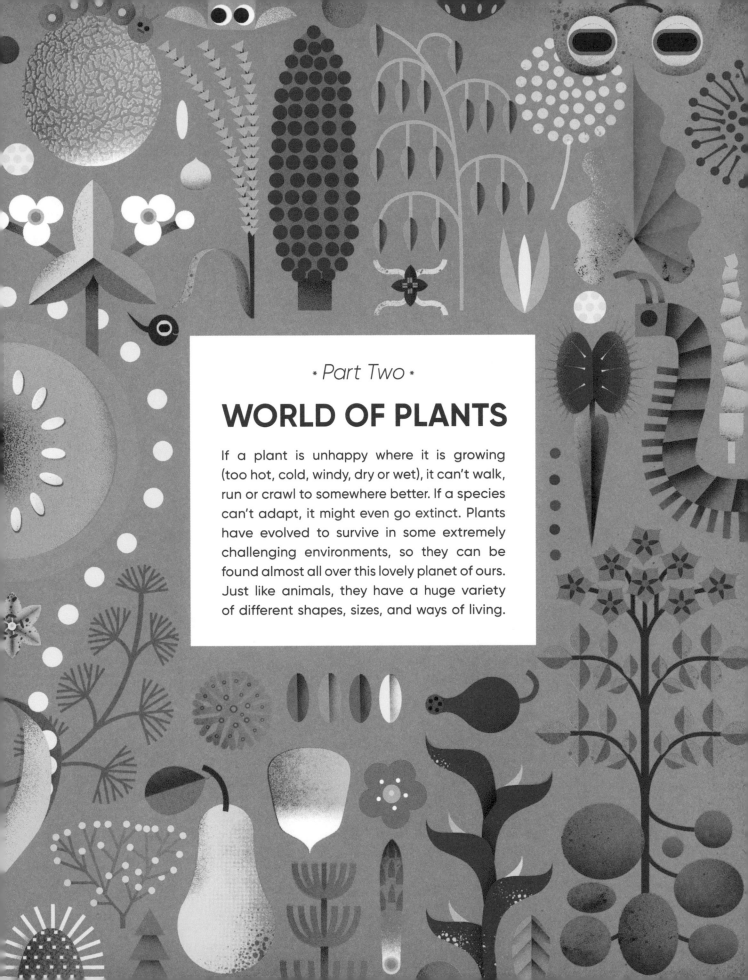

* Part Two *

WORLD OF PLANTS

If a plant is unhappy where it is growing (too hot, cold, windy, dry or wet), it can't walk, run or crawl to somewhere better. If a species can't adapt, it might even go extinct. Plants have evolved to survive in some extremely challenging environments, so they can be found almost all over this lovely planet of ours. Just like animals, they have a huge variety of different shapes, sizes, and ways of living.

THE PLANT KINGDOM

Every species of plant in the world originated from just one type of plant, millions of years ago. In the same way that animals have adapted, plants too have developed to be just right for their environment, through a process of trial and error known as **evolution** (see pages 44-45).

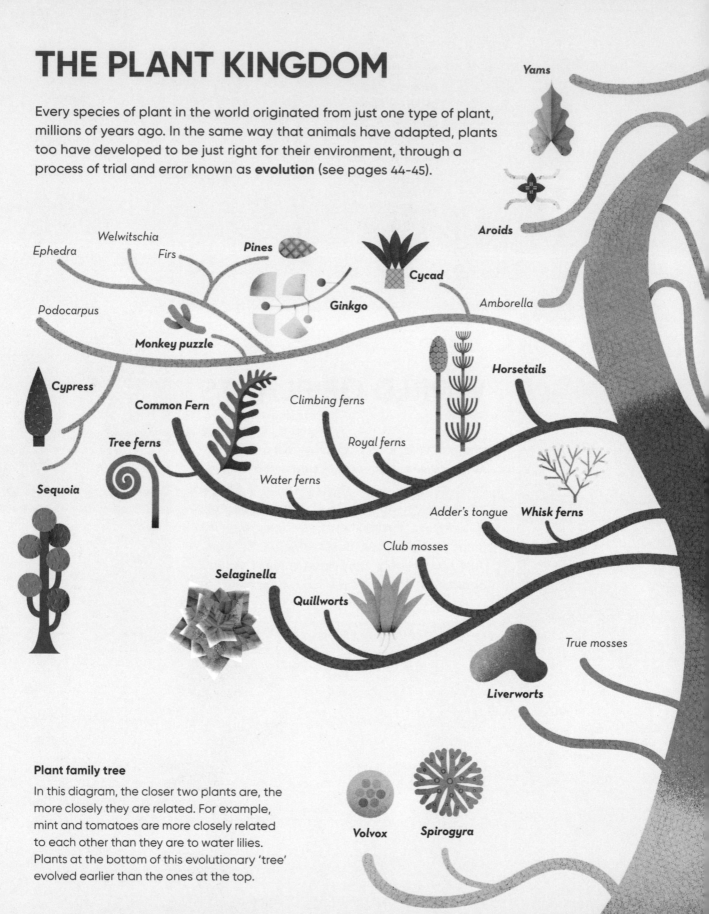

Yams

Aroids

Welwitschia

Ephedra

Firs

Pines

Ginkgo

Cycad

Amborella

Podocarpus

Monkey puzzle

Cypress

Common Fern

Climbing ferns

Horsetails

Tree ferns

Royal ferns

Water ferns

Sequoia

Adder's tongue

Whisk ferns

Club mosses

Selaginella

Quillworts

True mosses

Liverworts

Volvox

Spirogyra

Plant family tree

In this diagram, the closer two plants are, the more closely they are related. For example, mint and tomatoes are more closely related to each other than they are to water lilies. Plants at the bottom of this evolutionary 'tree' evolved earlier than the ones at the top.

38

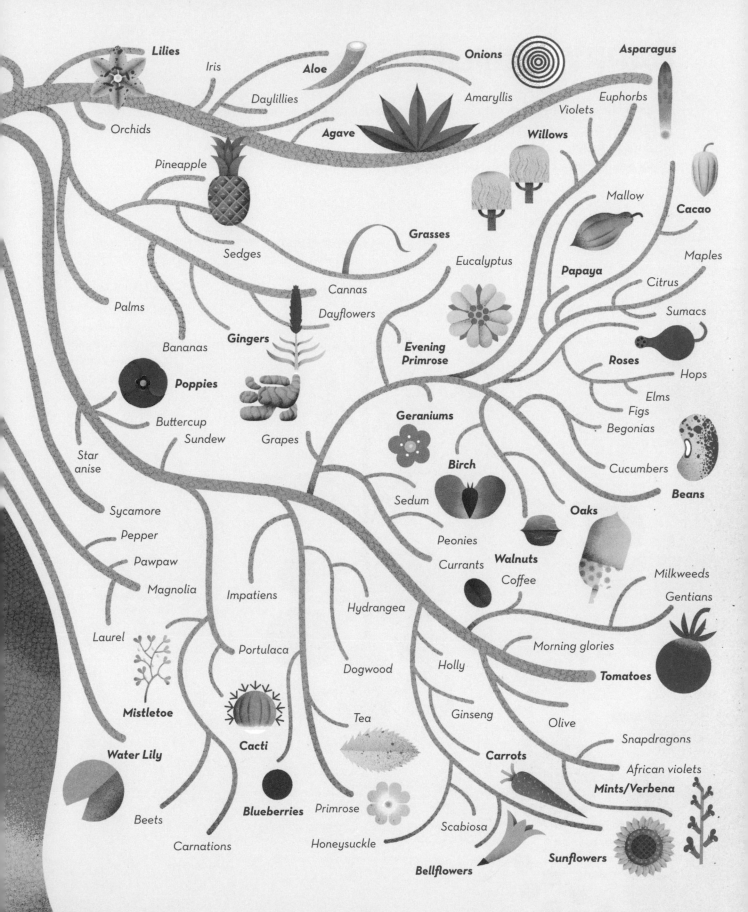

Lilies

Iris

Aloe

Onions

Asparagus

Daylillies

Amaryllis

Euphorbs

Orchids

Violets

Agave

Willows

Pineapple

Mallow

Cacao

Sedges

Grasses

Papaya

Maples

Eucalyptus

Citrus

Cannas

Sumacs

Palms

Dayflowers

Roses

Bananas

Gingers

Evening Primrose

Hops

Elms

Poppies

Figs

Buttercup

Geraniums

Begonias

Sundew

Grapes

Cucumbers

Star anise

Birch

Beans

Sycamore

Sedum

Oaks

Pepper

Peonies

Pawpaw

Walnuts

Milkweeds

Currants

Magnolia

Impatiens

Coffee

Gentians

Hydrangea

Laurel

Morning glories

Portulaca

Dogwood

Holly

Tomatoes

Mistletoe

Cacti

Tea

Ginseng

Olive

Snapdragons

Water Lily

Carrots

African violets

Beets

Blueberries

Primrose

Mints/Verbena

Carnations

Honeysuckle

Scabiosa

Sunflowers

Bellflowers

HAPPY FAMILIES

Taxonomists are scientists that group living things into categories. They have put plants in families based on how similar they are. There are plants that you are probably very familiar with (and use in your everyday life) but don't know are actually close relatives.

The Rose Family

The rose family (*Rosaceae*) contains roses, apples, pears, peaches, plums, nectarines, almonds, strawberries, sloes, raspberries, quinces, cherries, apricots and more.

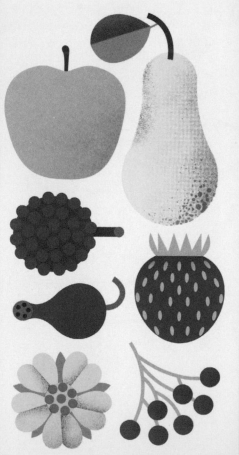

The Mint Family

The mint family (*Lamiaceae*) has mint, basil, rosemary, thyme, lemon balm, marjoram, oregano, lavender and others.

The Potato Family

The potato family (*Solanaceae*) contains potatoes, tomatoes, aubergines, tobacco, deadly nightshade, henbane, mandrake and chillies amongst others.

The Cucumber Family

The cucumber family (*Cucurbitaceae*) includes cucumbers, courgettes, gherkins, melons, squashes, luffa, pumpkins, and gourds.

The Bean Family

The bean family (*Fabaceae*) has French, runner, string, soya, broad, butter and Lima beans, peas, peanuts, lentils, tamarind and many more.

The Grass Family

The grass family (*Poaceae*), contains grasses, sugar cane, rice, oats, wheat, barley and cereals. Another grassy cousin is maize (also known as sweetcorn).

DIY: CORNFLOUR SLIME

Maize is an extremely versatile member of the grass family. In this activity, we will use it to make a-maize-ing slime! Cornflour slime is a special type of fluid that doesn't follow the usual rules.

You Will Need:

- ☐ Some fine cornflour (250g)
- ☐ Water (300ml)
- ☐ Some food colouring
- ☐ A wooden spoon
- ☐ A bowl

Natural Food colourings:

Turmeric = Yellow

Beetroot powder = Purple

Spinach powder = Green

Safety Note: If you have asthma, wear a face mask whilst pouring the cornflour. as the dust could irritate your lungs.

How to Make Your Slime:

1. Pour the fine cornflour into the bowl.

2. Gradually pour water into the bowl and mix until it looks and feels like custard.

3. Add about 15 drops of liquid food colouring or a teaspoon of natural food colouring and stir.

4. Now hit the mixture with your fist! It should become thicker. That's because slime is a special type of fluid. Unlike most runny things, when pressure is applied to slime, it gets more **viscous** (thick).

5. Make your natural slime last longer by keeping it in an airtight container in the fridge.

EVOLUTION

Just like us and other animals, plants come in different shapes and sizes and have developed ways to survive in different and challenging environments all over the planet. This variety has come about via a slow process called evolution.

Charles Darwin

The theory of 'evolution by natural selection' was first brought to the public's attention in the 19th century by the **naturalist** Charles Darwin. Evolution is the process by which organisms (living things) change over the generations, in how they look or how they behave. Organisms inherit these changes from their parents, for example, you might have inherited your eye or hair colour from your mum or dad. Darwin was fascinated with all kinds of living creatures, and studied everything from carnivorous plants to earthworms.

44

Bird brains

When he visited the Galápagos islands in the Pacific Ocean in 1835, Darwin studied the size and shape of finch beaks. He theorised that the finches' beaks had all adapted from one original species to suit different food sources. The finches changed from insect-eaters to nut-crackers, flower-eaters and even tool-users. Some of these birds were even using tools made of plants, just like we do.

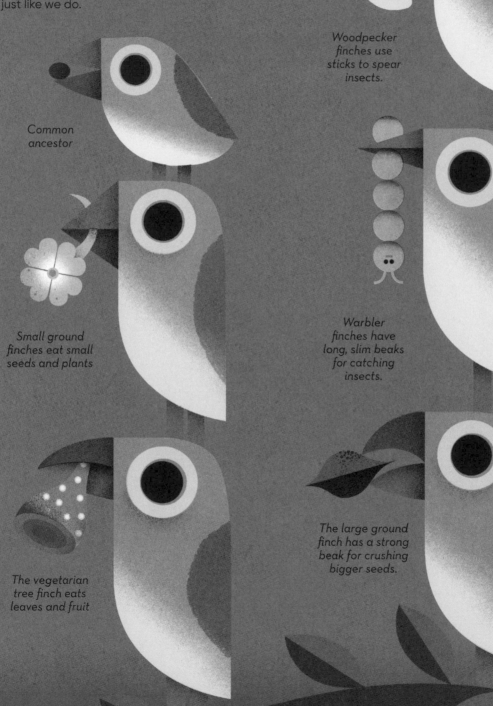

Woodpecker finches use sticks to spear insects.

Common ancestor

Small ground finches eat small seeds and plants

Warbler finches have long, slim beaks for catching insects.

The vegetarian tree finch eats leaves and fruit

The large ground finch has a strong beak for crushing bigger seeds.

ADAPTATION

Plants can die because they are eaten, don't get enough light, dry out or get a disease. Thanks to evolution, plants are **adapted** to suit their environments, which gives them a greater chance of surviving long enough to produce seeds. They then pass on these adaptations to their seeds.

A landscape from about 250 million years ago

Horsetails were the among the first land plants to evolve.

Survivors

Over millions of years, living things that have managed to reproduce have passed on their traits. Others didn't have the right traits for their environments and died out. Every change in the environment has presented new challenges that meant different life-forms survived or died. Through this process, living things changed from simple, single-celled plants and animals into the diverse array of life that we see in the fossil record and around us today.

DID YOU KNOW?

Around 252 million years ago, 90-95% of all life on Earth was wiped out in an event called The Great Dying. It is thought that volcanic ash blotted out the Sun. We don't know how, but some plants survived, including the ancestors of flowering plants.

LIFE IN EXTREMES: HOT AND DRY

In the desert, where it rains less than 25cm per year, plants have to either be good at growing with very little water or able to store it. Desert plants that are swollen with water in their stems or leaves are called 'succulents' and some of them can go for up to 100 days without a drink.

Agave

Prickly customers

The most famous example of a succulent is the cactus, originally from North, Central and South America. Their thick, green, fleshy stems are swollen with water and the leaves have been reduced to prickly spines. These spines are usually white and sometimes very densely packed to reflect sunlight – like a natural sunscreen – to cool the plant and stop it from losing precious water.

Deep roots

Another way of coping with such a small amount of water is to send roots all the way down to the groundwater far below – too far for most other plants. Some desert plants have been found to have roots that are 50 metres long, while others spread their roots outwards, close to the surface so that when it does eventually rain they catch as much water as possible and store it.

48

Night air

Every time a plant needs to breathe CO_2 in, it has to open its pores, called 'stomata'. If it is in a very hot place, it risks losing valuable water, so many desert plants do their 'breathing in' at night and store the CO_2 to be used in photosynthesis the following day.
Try breathing in and out with your mouth wide open for one minute and see how dry your mouth becomes.

Saguaro cactus

In the shadows

One way of keeping cooler in the desert is to live in the shadow of a taller plant. These helpful shade-givers are called 'nurse plants'. In the northern hemisphere, the north side of a plant receives less sunlight, which makes it a good place to hide from intense rays.

DID YOU KNOW?

Desert-dwelling people can survive finding underground water in desert plants with deep roots. They also locate plants with fleshy tubers that they can dig up and squeeze the juice out of.

Resurrection fern dry (left) and hydrated (below)

Plant zombies

Resurrection plants live in hot, dry places and look dead and crispy for most of the year as soon as the rain falls, they rehydrate and rejuvenate – many of them flowering and producing seeds in a matter of hours or days.
One example is the Rose of Jericho, or "resurrection fern" (Selaginella lepidophylla). It is a type of fern that rapidly opens up when put into a saucer of water.

49

LIFE IN EXTREMES: THE JUNGLE

If you live in an equatorial jungle, the sun will rise at around 6am and set at around 6pm every day and the annual rainfall is more than 200-225cm. There are no extreme seasons in the same way that someone living further north or south of the equator would experience.

Social climbers

There are so many plants fighting for light, water and space that some have evolved to grow on (epiphytes) or climb up (climbers) other plants. Epiphytes hold on tightly to their tree and may have dangling roots to take moisture from the air. Climbers will wrap themselves around a tall tree and reach up towards the sunlight.

Special leaves

Plants that live in very rainy places can encounter problems – if their leaves get too wet they will go mouldy, and if a raindrop stays on a leaf when the sun comes out, it can act like a magnifying glass and burn it. To combat this, some plants have leaves designed to divert water away. Many tropical leaves are waxy with gutters down their centres and pointed 'drip tips' at their ends.

Because of this constant warmth, frequent rain and regular amount of sunshine, the plants here thrive and some of the most biodiverse **ecosystems** on Earth have evolved. Rainforests are potential sources of many new foods and medicines, but they are quickly being destroyed for farming, fuel, roads and urban developments. We might be losing important plant resources that we don't even know about yet. It is crucial to **preserve** and **conserve** them while we still can.

Dark dwellers

It can be very dark on the forest floor, so some ground-dwelling plants like begonias have a special trick – they have reddish or purplish undersides, which act like mirrors. Sunlight bounces off the red part back up through the green leaf to give the chloroplasts a double dose of sunshine.

Begonia

Supporting roots

Rainforest soil is particularly thin and poor in nutrients because of the constant rainfall. Many plants develop 'buttress' or 'stilt' roots. These thick, wide roots stop the plant from falling over, and ensure they can absorb enough nutrients and minerals from the soil.

51

WATERY WORLD

It is thought that all plants evolved from water plants and, over millions of years, became more and more complex. Plants growing in water already have a bit of an advantage – they started off there. An advantage of life in water is that since water warms up and cools down slowly compared with air, aquatic plants are safe from the rapid temperature changes that land plants experience between day and night. The leaves and stems of such plants absorb minerals and water from their surroundings. Their roots anchor them to the bottom of lakes, ponds, rivers and oceans.

Water lettuce

Water hyacinth

Arrowhead

Yellow water-lily

Bladderwort

Water-crowfoot

Floaters

The red wavelengths found in sunlight are very useful to plants, but water filters it out, leaving only blue-green wavelengths. To allow them to float in the water and get them closer to the surface where there's more light, aquatic plants may have air bladders (as seen in some seaweeds) or stems with air pockets in them.

PLANT TRICKS

Temperate areas of our planet are found north and south of the tropical zones. They have four seasons: spring, summer, autumn and winter. Even in habitats such as woodlands, forests or fields where plants have space and light, they need a few special tricks to come out on top.

Efficient arrangements

Many plants, including sunflowers (*Helianthus*) arrange their leaves strategically around their tall stem, so that no leaf is growing directly above another. This helps temperate plants make the most of their sunlight, especially when seasons change, and days are shorter.

Sunflower

Grass survivors

About 40% of the world's land is grassland. Hungry herbivores graze constantly on grass, and so these plants evolved to be able to regenerate from deep, wide root systems. They can even grow back after fires! They have flexible stems, so they don't snap in windy weather.

Grass

54

Mistletoe

Hug of death

Strangler figs start by growing on a victim tree, then send a root down into the ground before growing back up and all around the tree. This acts as a support to begin with but then crushes the tree to death. Once the tree is dead and decayed, the fig lives on with a hollow space in the centre where its victim used to be.

Thieving plants

Dodder *(Cuscuta)* and mistletoe *(Viscum album)* are examples of plant parasites, meaning they invade other plants and steal the sugary sap that their host plant has made during photosynthesis. Dodder doesn't even have chlorophyll, and is really just an exploring stem and root system.

Strangler fig

DIY: EVERGREEN FREEZE

This brrr-illiant scientific test will show the effects of freezing conditions on deciduous and evergreen trees.

Deciduous trees include maple, oak, lime, poplar and beech.

Evergreen trees include fir, pine and cypress.

You Will Need:

☐ Two small fallen deciduous branches with leaves

☐ Two small fallen evergreen branches with needles

☐ A freezer

How to Make Your Freeze:

1. Take one of the deciduous branches you have collected, and one of the evergreen branches, and place in the freezer for a few hours.

2. Leave the other deliciduous and evergreen branches somewhere at room temperature for the same amount of time.

3. Remove the branches from the freezer. What do you notice? Once the branches thaw, you will see that, whilst the evergreen leaves will look much the same, the deciduous leaves have changed.

The evergreen leaves are much tougher and are adapted to withstand very cold conditions.

The deciduous leaves will have gone dark and floppy. The cells inside have burst.

POISON

There are about 428,000 species of plant in the world, but experts believe that only 5% are actually **edible**. Why are the majority of plants **poisonous**?

Hemlock

Foxglove

Deadly nightshade

Lily of the valley

Poison protection

Unlike animals, plants can't run away if they are being attacked. For protection, they may have spikes or hairs to make themselves difficult or unpleasant to eat. Most plants protect themselves in invisible ways: they contain chemicals that taste bitter, spicy or sour... some can even kill!

58

Greater grub

A plant that is edible is safe to eat and won't make you sick. However, that doesn't mean it tastes nice! The foods we buy in shops are the result of thousands of years of selective breeding by farmers to bring out the best flavour and nutrition for our bodies. The wild relatives of these plants, which our ancestors ate long ago, probably didn't taste that great at all.

Ancient Mexican farmers began to select better maize plants many thousands of years ago. Over the years, maize cobs became bigger, fuller, sweeter and tastier.

Castor oil plant

Castor oil seeds

Help or harm?

The castor oil plant (*Ricinus communis*) contains a useful oil which can soothe eczema or help your digestion. Its seed coat, however, contains a deadly chemical called ricin, which is one of the most poisonous substances in the world. It can kill a person without leaving much of a trace. Stay well away from this plant if you ever encounter it!

59

FOOD CHAINS AND FOOD WEBS

Humans (and other animals) can't get energy directly from the sun, so we have to eat plants or other animals to survive. The sunlight energy flows through a food chain, from plants to plant-eaters and then to the meat-eating predators that hunt them and are hunted in turn.

Most food chains don't have more than four or five stages because energy is lost at each level in the form of heat (which we lose through our skin) and poo.

When lots of other chains join together, we have an entire **food web**! Take a look at this diagram to see how this works.

CARNIVORES

HERBIVORES

PLANTS

SUNLIGHT

Ladybird

Hedgehog

Snake

Dragonfly

Wasp

Aphid

Moth

Butterfly

Squirrel

Eagle

Fox

Swallow

Beetle

Hummingbird

Mouse

Bee

Frog

Grasshopper

Caterpillar

Owl

Blue tit

Snail

Rabbit

Each stage in a food web is called a 'trophic level'. The word trophic comes from the Greek word *trophe*, meaning 'nourishment'.

IT'S A TRAP!

Plants living on the edges of rivers and in marshes and bogs tend to have their nutrients washed away from their roots. To make up for this loss, some have evolved the ability to lure and catch insects. These are usually known as **carnivorous** or insectivorous plants. You could think of them as vampires since they don't actually eat their prey – they drink them instead! There are two types of plants that do this: those that move (active traps) and those that don't move (passive traps).

Venus flytrap

Sundew

Active traps

Perhaps the most famous active trap plant is the Venus flytrap *(Dionaea muscipula)* which has a leaf folded like the pages of a book, lined with tasty nectar. They are also lined with sensitive trigger hairs which can tell when a fly or other insect has landed on it. This triggers the leaf to quickly snap shut and trap the insect, before filling its closed leaf with digestive juices.

A sticky ending

Sundews *(Drosera)* have sticky leaves which look like they are covered in little drops of water. When a fly lands on a 'droplet' thinking it's going to get a drink, it realises that these drops are actually very gluey. Unable to leave, the fly dies of starvation and the leaf slowly wraps itself around the fly to suck the juices out of it. Yuck!

Passive traps

North American pitcher plants (Sarracenia) are passive trap plants with leaves shaped like empty ice-cream cones. Around the top of the leaves is attractive nectar, which entices flies and other insects over for a snack. But this is no ordinary nectar – it is in fact a sleeping potion. The edges of the plant are smooth and waxy. Sleepy insects slip and fall inside the leaf and cannot get out, as the walls are covered in downward pointing hairs that make it really difficult to escape. To make matters worse, there is digestive liquid at the bottom in which the exhausted insects drown. Next, the insects partly dissolve, and as other insects fall in too, they all mush together into a delicious and nutritious insect smoothie.

Pitcher plants

63

* Part Three *

FROM BREAKFAST UNTIL BEDTIME

From the moment we wake up to the moment we fall asleep, plants sustain us. From essential activities like eating and breathing right through to the way we choose to spend our free time, if you look closely enough, you'll discover we are surrounded by plants.

I ATE SUNSHINE FOR BREAKFAST – AND SO DID YOU!

While you eat your breakfast, millions of other meals are taking place in your garden, street, neighbourhood and even down the back of your sofa. From microbes through to invertebrates, small mammals and birds – everyone has to eat something.

If you happened to eat a bowl of breakfast cereal this morning, then you basically ate a whole load of mashed up grass seeds – perhaps maize, wheat, oats, rice and rye. Most of them include some sugar, either from sugarcane (*Saccharum officinarum*) or sugar beet (*Beta vulgaris*).

You may have had some toast with butter and jam on it for your morning meal. If so, you ate some wheat, (or maybe rye, oat or barley, depending on what bread you like), strawberries or raspberries, sugar and oil, from either plants or animals. In food chain form, your breakfast would look like this:

Sunlight energy

Wheat plants

Wheat processed and cooked into bread

You!

PLANT JUICE

Did you have a drink with your breakfast this morning?
Most drinks are made from plants. Here are a few examples:

Tea

People have been drinking tea for at
least 5,000 years. It is now the most
popular beverage in the world after
water, with the tea-loving people
of Turkey drinking more per person
than in any other country. In China,
offering a cup of tea shows respect
to an elder, says sorry or gives thanks
on a wedding day. Tea is great for
your body: it contains antioxidants,
helps to control your cholesterol, and
speeds up your metabolism. There are
an estimated 1,500 different types.

Coffee

According to legend, a group of
Ethiopian goat herders were tending
their flock of goats one day and
noticed the animals were eating
some unfamiliar seeds. That night,
the goats were full of energy and
didn't sleep. When the goat herders
picked and tried the beans, it had a
stimulating effect on them too, and
so coffee was 'born'. This drink from
Coffea arabica and other species is
extremely popular because its high
caffeine content wakes us up.

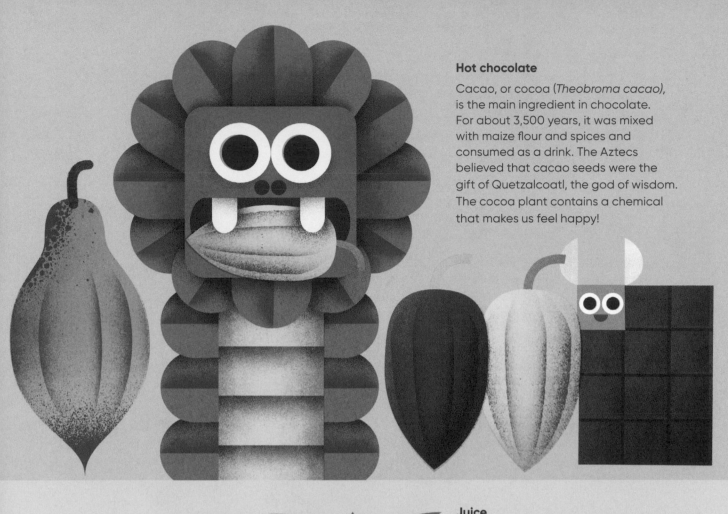

Hot chocolate

Cacao, or cocoa (*Theobroma cacao*), is the main ingredient in chocolate. For about 3,500 years, it was mixed with maize flour and spices and consumed as a drink. The Aztecs believed that cacao seeds were the gift of Quetzalcoatl, the god of wisdom. The cocoa plant contains a chemical that makes us feel happy!

Juice

Containing tons of nutrients, juices are packed with goodness. They can help protect your immune system, cells and organs because they contain useful vitamins. Orange, apple, grape, cranberry, grapefruit and tomato make some of the most popular juices, but you can also use vegetables such as beetroot, watercress and celery, and add herbs and spices such as parsley and ginger.

TIME TO BRUSH YOUR TEETH

Toothpaste contains several plant **extracts**. Many are mint-flavoured, and most contain some wood pulp (yes that is mashed up trees) called cellulose gum to hold the ingredients together. Corn starch (made from ground maize seeds) is often used for this too.

Brush with bamboo

It's estimated the average person will use 300 toothbrushes in their lifetime. Unfortunately, a lot of these will end up in the ocean or landfill, and the plastic they are made from can take up to 1,000 years to disintegrate! Bamboo toothbrushes are a much more eco-friendly alternative, and are made using natural, biodegradable materials including bamboo wood, charcoal and castor oil.

Twiggy teeth

Throughout the Middle East and parts of Africa, many people brush their teeth with twigs called miswaks, which are made from *Salvadora persica* (known as the toothbrush tree). This is a rough plant which contains chemicals that kill germs, as well as the substance fluoride, which protects tooth enamel.

Botanical companions

Look at yourself in the bathroom mirror. What is your name?
You may know or even be a Lily, Poppy, Ivy, Olive, Daphne, Iris,
Rosemary, Rose, or perhaps a Petunia, Myrtle, Marigold, Marguerite,
Blossom, Basil, Bryony, Holly, Hazel, Heather, Woody, Rowan,
Primrose, Tulip, Petal, Jasmine, Ginger, Apple, Hyacinth, Tamarai
('Lotus flower' in Tamil), a sweet William, Violet or Daisy!

CLEAN UP!

When you do the washing up, you'll probably be using a plastic brush to scrub the dishes. But before we could make these out of plastic, people used a plant called butcher's broom (*Ruscus aculeatus*) which is tough and spiky and perfect for scrubbing. Another useful scrubbing plant is called horsetail (species of *Equisetum*). There is even a species of horsetail called the scouring rush.

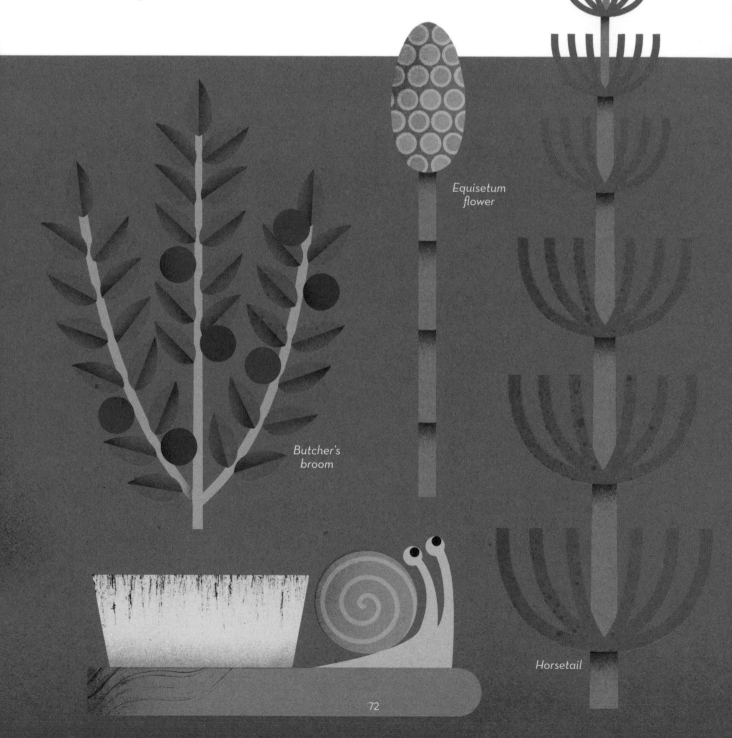

Equisetum flower

Butcher's broom

Horsetail

The lotus effect

The lotus plant (*Nelumbo nucifera*) is a sacred and symbolic plant in Buddhism and Hinduism. It grows in murky waters but comes out sparkling, with a beautiful pinkish flower and pristine leaves. The lotus has a slippery secret – its leaves are covered with lots of microscopic bumps. These bumps make it difficult for dirt to stick, and so the dirt clings to droplets of water instead and rolls right off. It's a surface that cleans itself!

Close-up view of a water droplet on a lotus leaf

The lotus plant has inspired scientists to invent self-cleaning paint for the outside of buildings, and glass and plastics that never get wet and foggy, which are used for cars and helmet visors. This is a great example of biomimicry, where science copies something from the natural world.

DIY: SHELF LIFE PROJECT

We throw away lots of things every day that can be used again or regrown. Try to collect seeds from the food you eat, and jars, bottles, tins and containers that would otherwise be thrown away to create an upcycled garden. We call this idea the 'Shelf Life Project'. It was developed at the Chelsea Physic Garden in London, UK, but is intended to be used all over the world.

How to Grow Your Upcycled Garden:

Peanuts

These will grow as long as they are not roasted. You'll need to keep them warm and remember that the peanuts are produced underneath the soil. Check under the soil after four months to check their progress.

Potatoes

Fill an empty crisp packet with some grit then some compost, until it's one third full. Add a smallish potato (or even some peel), and then fill almost to the top with more soil. Keep the soil moist and see what happens!

Ginger

Fresh root ginger grows easily indoors in the warmth. Half-bury the root in sandy compost and keep moist. You'll need to use a wide pot as ginger plants like to grow their roots sideways.

You Will Need:

- Empty jars, bottles and packets (use some scissors to make drainage holes in the bottom of these)
- Some water
- Compost
- Clingfilm
- Anything that can be replanted! See below for some examples.

Top tip: Fill the containers with a little grit or gravel. This will weigh them down to stop them from toppling over, and add drainage.

Safety Note: Be careful of sharp edges on tin cans! Cover them with some insulating tape for safety.

Citrus fruits
Fill a jar three quarters of the way with moist compost. Place seeds of lemons, oranges, satsumas, or grapefruit on the surface. Cover with clingfilm and leave in a warm place. After a few weeks, shoots should appear.

Avocado
Soak the stone overnight. Using a jar, submerge the lower half of the stone in fresh water, sticking in toothpicks to support it. Change the water regularly and see what happens.

Tomatoes
Get seeds either directly from the fruit or from a garden centre. A dwarf variety might be best, as the plant can get quite big. Tomatoes grow well in a tin, or an empty carton – or even a used-up bottle of tomato sauce!

LET'S GET DRESSED

Now that you've washed and stared at yourself in the mirror, isn't it time to get dressed? The thread-like fibres of many plants have been used to make clothes for thousands of years. After being harvested, they are spun into longer threads and woven into fabrics.

Wearing clothes for fashion, instead of just for keeping warm, is thought to have occurred early on in human history. The fine fibres of the flax plant (*Linum usitatissimum*) were woven into clothes and were even used as burial shrouds for the Egyptian pharaohs from around 3000 BCE.

DID YOU KNOW?
Sweet-smelling lavender (Lavandula angustifolia) was used in mummification to make the musty mummies smell nicer. Lavender is a mild sedative, so it is good for helping people get to sleep. Lavender oil can also soothe minor burns.

Bamboo
(Bambusoideae)

Bark and bamboo

Cloth can be made from tree bark, like that of the paper mulberry (*Broussonetia papyrifera*) and a type of Ugandan fig tree (species of *Ficus*). Fabric that is made from bamboo is very soft. It allows the skin to 'breathe' and not get too sweaty, so is perfect for making socks for people with smelly feet!

Cotton

The fine threads of this plant's fruits, or 'bolls', are individual cells that are very, very long. It is used in textiles, chewing gum, bank notes and paper, and the oil of its seeds is used for cooking and in products like soap, candles, margarine and plastics.

Cotton
(Gossypium)

Coconut

This fruit and other palm fibres can be turned into string or rope. Raffia palm (species of *Raphia*) has very long leaves and can be used for mats, baskets, shoes, clothes and hats.

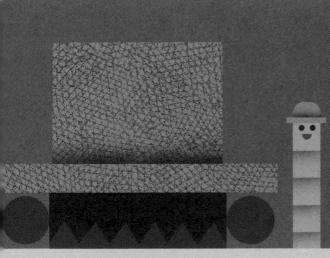

Coconut
(Cocos nucifera)

Stinging nettle

Stinging nettle

This spiky nettle (*Urtica dioica*) and hemp (*Cannabis sativa*) are both plants that can be used for paper- and fabric-making, as well as in medicines.

77

SWEET AND SOUR SMELLS

After you've woken up on cotton sheets, you might head into the bathroom to wash with plant-based soaps and shampoo. The earliest recorded types of soap would have been made from chemicals called saponins found in some plants (it can also be sourced from animals). Many modern-day soaps contain olive oil.

Lather up

Some plants have natural soapy properties, like soapwort (*Saponaria officinalis*), soap bulb or yucca plants. These can be chopped up and lathered in your hands. Soaps usually contain a lot of vegetable oil to keep skin soft and supple, using oils such as olive, coconut, macadamia nut and avocado.

Soap bulb plant

Soapwort

Heady scents

For thousands of years, people have burned scented wood called incense, such as Frankincense (*Boswellia spp.*) in religious ceremonies. People have also crushed flowers to make perfume. Essential oils used in perfume are often very expensive – it takes about 2.5 tonnes of rose petals to make one teaspoon of rose oil.

78

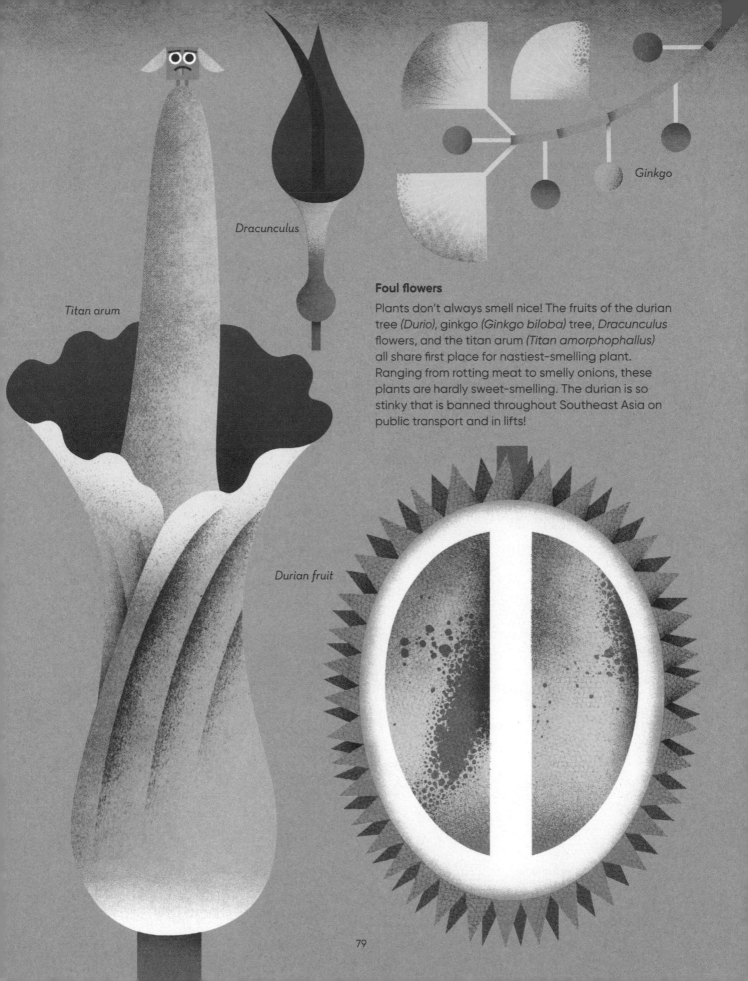

Titan arum

Dracunculus

Ginkgo

Foul flowers

Plants don't always smell nice! The fruits of the durian tree *(Durio)*, ginkgo *(Ginkgo biloba)* tree, *Dracunculus* flowers, and the titan arum *(Titan amorphophallus)* all share first place for nastiest-smelling plant. Ranging from rotting meat to smelly onions, these plants are hardly sweet-smelling. The durian is so stinky that is banned throughout Southeast Asia on public transport and in lifts!

Durian fruit

COLOURFUL WORLD

Imagine a world without the bright and vibrant colours we have on our clothes and bags, curtains and walls. Early humans would have worn clothes and owned objects that were mostly grey and brown. The discovery of pigments from rocks and natural dyes from plants and animals revolutionised how the world around our ancestors looked.

Making paints

Until human-made pigments were invented in the 18th century, all paints and dyes came from the natural world. From prehistoric times, people made colourful pastes from ground-up rocks and plants. These paints included bright blues, vibrant reds, rich purples and vivid yellows. A deep black can be made from peach or cherry stones.

Smart art

Poppy, walnut, safflower and linseed oils are all used by oil painters to help paint to dry more quickly or slowly or to create different sheens on the final painting. Starch from potatoes can also make gouache paints creamier.

Making your mark

Charcoal is made by burning wood. It is often made from the twigs of willow trees (a species of *Salix*) and grapevines (a species of *Vitus*) and is great to draw with as it produces bold, black strokes and can be nicely smudged for more 'blurry' effects.

Colourful clothing

To make our clothes look more interesting, they are dyed different colours – with synthetic or natural dyes – including some vibrant exotic blues, reds, purples, yellows and greens using plants such as indigo (*Indigofera*), woad (*Isatis tinctoria*), madder (*Rubia tinctoria*). Everyday things like onion skins, rhubarb and turmeric work too.

Rhubarb

Woad

Turmeric

Madder

Royal blue

The most vibrant colours were very expensive before artificial dyes were invented. As only the wealthy could afford them, they became status symbols. Indigo blue was the colour of royalty in France until an artificial blue dye was invented in the 1890s and it became less exclusive.

DIY: LEAF PRINTING

Use this technique to decorate bookmarks made from coloured card. Give them as gifts to friends and family or use one to keep your place in this book!

You Will Need:

☐ A pair of rubber gloves

☐ At least one ink stamp pad in a dark colour

☐ Some plain, pale-coloured paper or card

☐ Some fresh leaves (cooking sage works well, but any leaf with bumpy underside veins works)

How to Make Your Print:

1. Put on your gloves. Then, place your leaf on to the ink pad and place some scrap paper over the top.

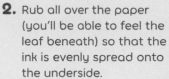

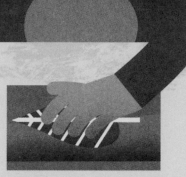

2. Rub all over the paper (you'll be able to feel the leaf beneath) so that the ink is evenly spread onto the underside.

3. Take the leaf and carefully place it on to the paper or card you want to print on to.

4. Put another piece of scrap paper over the top and rub all over again to release the ink.

5. You'll have a fine impression of that leaf and can repeat it to make a pattern!

As long as 10,000 years ago, Japanese houses were thatched with straw to keep them dry,

Traditional Cameroonian Baka houses are woven from thin sticks and covered with Ngongo (Megaphrynium macrostachyum) leaves.

Iron Age (1200-600 BCE) European roundhouses were built from strong wood such as oak, and thatched with grass.

Traditional Dinka houses in South Sudan are raised on tree branches to avoid floods.

AROUND THE HOUSE

In the distant past, humans wandered from place to place gathering food. When humans started farming about 10,000 years ago, they needed to stay in one place so that they could take care of their crops. People settled down, forming villages and towns. They started to build houses using plant materials like wood, bamboo, leaves, grass and straw, held together by soil, clay, mud, and even poo!

Resilient roofing

To keep the wind and rain out and the heat in, a house needs a roof. In tropical places, people have traditionally used palm and banana leaves for roofing materials. From the very beginning of housebuilding, roofs have been made by 'thatching', where water reeds (*Phragmites australis*) or other similar plants are tightly packed to form a thick and weatherproof surface. Plants such as bracken ferns (species of *Pteridium*) and mosses (*Bryophyta*) have also been used to cover shelters and keep warmth in.

A thatched cottage, 16th century England

Banana leaf

Palm leaf

The right wood

Wood is strong and lasts a very long time. Oak and maple wood are often used for walls, floors and ceilings. The world's oldest wooden building is the Horyuji Temple in Japan. It has been standing since the year 700. It is built from the wood of a type of cypress called hinoki (*Chamaecyparis obtusa*).

85

PENCIL AND PAPER

It's time to sit down and relax. You might choose an activity like writing a story or reading a book, which – you've guessed it –use materials that come from plants.

Writing materials

The pencils you use to draw and colour are made from incense cedar and juniper wood. They sometimes come tipped with erasers for rubbing out mistakes, which are often made from rubber. Although lots of stationery is now made from plastic, you can still find wooden rulers for drawing nice straight lines and measuring lengths. Box (*Buxus sempervirens*) is typically the wood of choice for these.

Making paper

Almost all of the paper we use in our everyday lives, from books and comics, to restaurant menus and cardboard boxes, is made from trees. Different kinds of trees produce different textures of paper. Softwoods like pine and birch have longer fibres and give paper more strength. Hardwood fibres are shorter but tend to work better for printing and writing papers. Pine, birch and Eucalyptus trees are also perfect to make paper as they are fast-growing species so it's easy to grow more of them!

1. A log goes into the pulping machine and gets cut up into wood pulp. Wood pulp is a watery mixture made up of wood fibres called cellulose, water and chemicals. Pulp is a bit like a woody soup!

2. Once the pulp is ready, it is sprayed onto a mesh screen to make an even, pulpy mat.

3. The dried paper is rolled onto huge rollers up to 10m (32ft) wide.

4. Shiny finishes and colourful inks are sprayed on, and the paper becomes a book or magazine.

DID YOU KNOW?

On average, one tree can be made into 8,335 sheets of A4 paper!

87

STRIKE UP THE BAND

The famous musical composer Felix Mendelssohn used to sit beneath a beech tree (*Fagus sylvatica*) at a place called Burnham Beeches in the UK. Much of his inspiration for writing orchestral pieces such as *A Midsummer Night's Dream* came from these woods. The stump of this tree can now be found the grounds of London's Barbican Centre in the UK. The instruments that are used to play these wonderful pieces of music are often made of wood or other plant products. Here are a few musical morsels:

Rosin, made from the sap of the pine tree, is rubbed on bows used to play stringed instruments like violins or cellos to get a better sound.

DID YOU KNOW?

The name "Xylophone" means "wood sound" in Greek! Can you guess what this instrument is made of?

In the past, the black keys on a piano were made from ebony. Ebony is a type of rich, dark hardwood. It is perfect for piano keys as it absorbs oil from the pianists skin and is strong enough for repeated tapping! Nowadays, most piano keys are made from plastic.

Bamboo is used to make wind instruments like flutes and pipes.

A grass called *Arundo donax* is used to make reeds (mouthpieces) for clarinets and saxophones.

Early gramophone records had a needle made out of the spines of a prickly pear cactus (species of *Opuntia*), and the actual record might have been made out of a kind of plastic that came from cotton.

DIY: GRASS SQUEAKER

You can make music with just a simple blade of grass! Here's how:

You Will Need:

☐ A blade of grass

☐ A pair of hands

How to Make Your Squeaker:

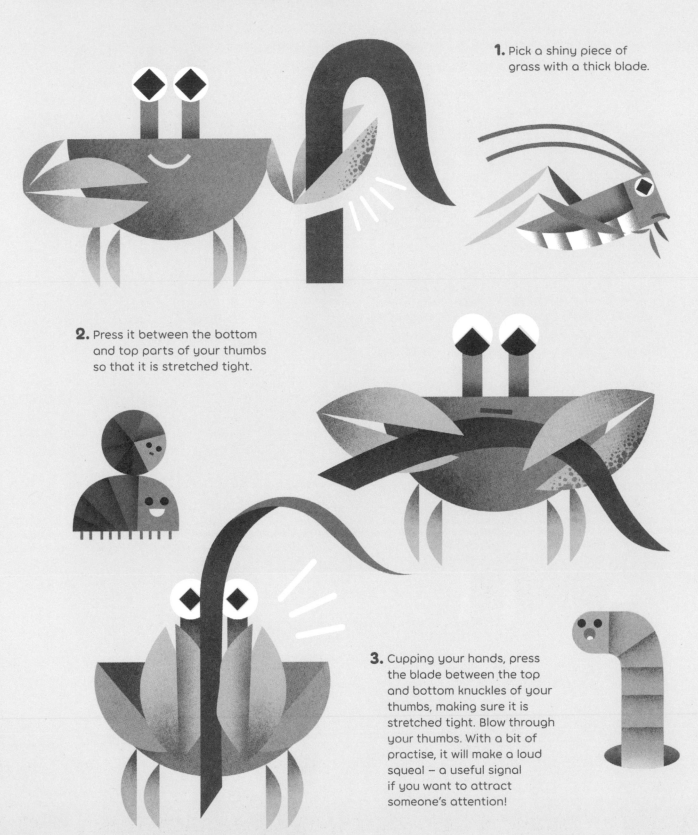

1. Pick a shiny piece of grass with a thick blade.

2. Press it between the bottom and top parts of your thumbs so that it is stretched tight.

3. Cupping your hands, press the blade between the top and bottom knuckles of your thumbs, making sure it is stretched tight. Blow through your thumbs. With a bit of practise, it will make a loud squeal – a useful signal if you want to attract someone's attention!

SPORTING LIFE

From simple ball games to highly athletic and competitive modern activities like tennis and football, people across the world love to play sports. Much of the equipment we still use to this day is plant-based.

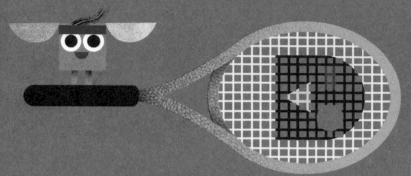

Making a racket

Old-fashioned tennis and squash rackets were made from trees like ash (*Fraxinus*), maple (*Acer*), American sycamore (*Platanus occidentalis*), hornbeam (*Carpinus betulus*), hickory (*Carya*), beech (*Fagus*), mahogany (*Swietenia*) and obeche (*Triplochiton scleroxylon*). These trees have also been used for sports equipment like hockey sticks, baseball bats and lacrosse sticks. This is mainly because the wood can take intense impacts without snapping – which is the same reason some trees are more flexible than others in high winds.

A spot of cricket

Cricket is another sport steeped in plants: cricket bats are made from the Cricket Bat Willow (*Salix alba caerulea*). The wood needs to be strong with balls hurtling at it at great speeds. The other parts of the bat feature rattan (*Calamus*) and rubber (*Hevea*). The ball is made from cork and wool all covered in leather. The stumps and bails are from the wood of the ash tree (*Fraxinus excelsior*).

Bouncing balls

Modern ball games still rely on rubber from rubber trees. Kit such as footballs, basketballs and tennis balls are all lined with rubber on the inside to keep air in and make them bounce.

Dangerous game

An early example of a game that relied on rubber was the "Mayan ball game". It was played with a large rubber ball, which had to be passed through a stone ring without the ball touching the player's hands or the ground. To add to the pressure, some historians believe the captain of the losing team was killed!

DIY: BEAN BAG BOULES

Back before the Olympics, the ancient Egyptians used to hold weightlifting competitions with bags full of grain (seeds such as wheat). On this page is a sport based along these lines and is very similar to the French game of boules or petanque. You can play on your own or with a friend.

You Will Need:

- ☐ At least one pair of old socks (no holes!) per person playing
- ☐ A bag of dried beans or peas
- ☐ An orange (not too ripe as you are going to be throwing it)
- ☐ An open, flat space, preferably outside
- ☐ String or a cable tie

Safety Note: Be mindful where you throw your boules! Never throw directly at someone else.

94

How to Make Your Bean Bag Boules:

1. Fill the socks with the beans or peas so that they reach just above the heel (less than half full). Make sure it's a fair game by weighing each bean-filled sock to be sure they weigh the same.

2. Making sure the beans are packed in tightly, twist the sock above the level of the beans and then use the string or cable tie to keep them in. You should end up with a bean-shaped, bean-filled sock! It might be spherical or a little bit more sausagey in shape.

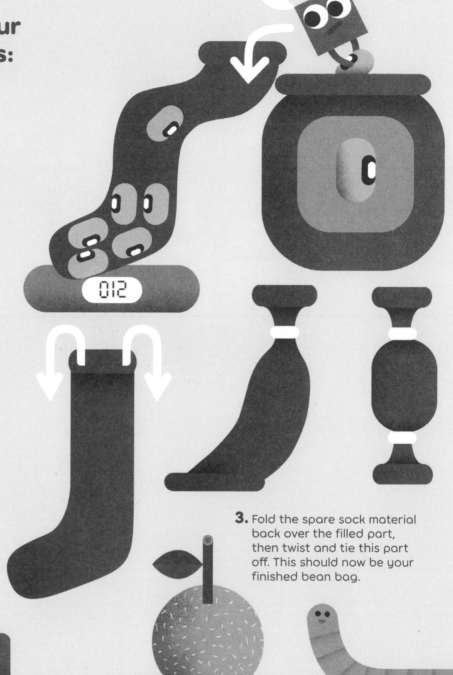

3. Fold the spare sock material back over the filled part, then twist and tie this part off. This should now be your finished bean bag.

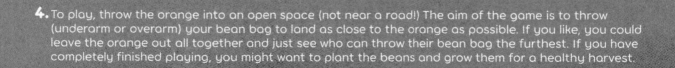

4. To play, throw the orange into an open space (not near a road!) The aim of the game is to throw (underarm or overarm) your bean bag to land as close to the orange as possible. If you like, you could leave the orange out all together and just see who can throw their bean bag the furthest. If you have completely finished playing, you might want to plant the beans and grow them for a healthy harvest.

95

* Part Four *

THE POWER OF PLANTS

Plants may be beautiful to look at, but their purpose on our planet is much more than this. Plants make our world go round. From the bank notes we use, to the medicine we take, plants are at the forefront of some of the most groundbreaking developments in science and technology. We still have a long way to go in discovering everything these green heroes can do.

SMART PLANT TECHNOLOGY

Science is defined as *"the knowledge about or study of the natural world based on facts learned through experiments and observation"*. Over thousands of years humans have done just that, through trial and error. Plants have been along for the ride as we observed and experimented with the world around us.

A bright idea

In the 1880s, inventor Lewis Howard Latimer was working with the famous inventor Thomas Edison, who had pioneered the electric light bulb. Latimer modified and improved Edison's design by replacing the part that lit up with a bamboo filament that lasted longer. Although this was eventually replaced by the metal tungsten and then by modern energy-saving bulbs, it was an important step in inventing efficent electric lights.

Mining for gold

They say money doesn't grow on trees – but gold might! Scientists have recently discovered that certain plants, such as Indian mustard (*Brassica juncea*), accumulate tiny silver and gold particles from the soil. These can then be harvested in a process known as **phytomining**, and used to make electronics.

Home-made torch

If you're lost outside in Polynesia on a dark night, the candlenut tree (*Aleurites moluccana*) from Samoa can double up as a handy torch! By threading its oily seeds together on a coconut leaf, the plant can be ignited and makes a very bright, long-lasting torch. The black soot that is left when the seeds burn is collected and used to make an ink used for tattoos.

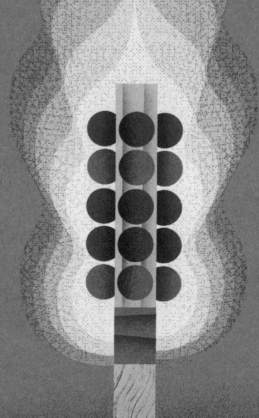

DIY: POTATO POWER PLANT

Did you know the humble potato is in fact an electric powerhouse? The mineral-rich water found in potatoes can conduct electricity. Used as a type of organic battery, potato power could be how we charge our household devices in the future. Try this experiment and find out for yourself.

You Will Need:

- A large potato
- A very small light bulb or LED light
- Two coins
- Two zinc-plated nails
- Three pieces of plastic-covered copper wire

Safety note: Be careful when handling the ends of the wires, because there is a small electric charge running through them.

100

How to Make Your Power Plant:

1. Cut the potato in half, then cut a small slit into each half, large enough to slide a penny inside.

2. Strip the plastic off the ends of the wires. Wrap some wire around each penny a few times. Use a different piece of wire for each penny.

3. Stick the pennies into the slits in the potato halves.

4. Wrap some of the third copper wire around one of the zinc-plated nails, then push the nail into one of the potato halves.

5. Take the wire connected to the penny in the half of potato, and wrap some of it around the second nail. Stick that second nail into the other potato half.

6. Connect the two loose ends of the copper wire to the light bulb to make it light up! Keep the wires as short possible so the electricity doesn't have far to travel.

HUNTING AND FIGHTING

As long as 10,000 years ago, many tools and weapons used by our ancestors were made from wood, such as clubs, spears, and boomerangs and kylies (types of throwing weapons). For thousands of years, armies around the world used wooden bows and arrows to fight enemies from afar, even on horseback.

Hunting their prey

Ancient people hunted their prey using sharp wooden spears. A spear made from the yew tree (*Taxus baccata*) was crafted and used about 420,000 years ago by an ancestor of modern humans, the species *Homo heidelbergensis*. To get the tip of the spear through a woolly mammoth's thick hide, the hunter had to get up very close to the huge animal.

Bows and arrows

Strong and flexible longbows made of yew were the most lethal weapon in Europe for about 300 years, between the 13th and 16th centuries. An arrow fired from this bow could slice into armour from 200 metres away. Sometimes the archers would even dip the arrows in the poisonous sap of the yew for extra deadly force.

Lighting the way

Cyathea dealbata, also known as the silver tree-fern, is a fern from New Zealand. Māori hunters would use it to find their way back home after hunting at night, as the undersides of these plants' fronds are visible in low light. This way, they didn't have to go out with flaming torches, which would scare animals away with their smoke and bright flames.

Stealth mode

If you want to blend into the scenery around you, what could be better than to use that very same environment? For soldiers trying to go unnoticed in natural environments, leaves, twigs and soil make great stealthy outfits. Modern, printed camouflage is still often brown and green to blend in with natural environments.

DIY: INVISIBLE INK

This fun task is useful for leaving very important and private messages for someone. Follow the steps and make up your own intriguing message.

You Will Need:

- [ ] A lemon or onion
- [ ] A small bowl
- [ ] A knife
- [ ] A nib pen
- [ ] A sheet of paper
- [ ] A lemon squeezer

Safety Note: Ask an adult to help you cut and squeeze your lemons or onions.

How to Make Your Ink:

2. Using the juice as an ink, with a clean nib write a message on a piece of paper.

1. Cut a lemon or onion in half. Squeeze the juice into a bowl using the lemon squeezer.

3. Leave it to dry. As the message dries it will disappear!

MEET ME AT THE TREEHOUSE

4. To make it reappear, hold the paper over a lamp or radiator. The heat will cause the invisible ink to darken so it can be read.

GREEN HEALING

It's good to eat a healthy, balanced diet and get enough exercise. This is called preventative medicine. Herbs and spices, which make our food tastier, also have important nutrients and act as little, regular doses of medicine. In fact, the 'father of modern medicine', the Ancient Greek Hippocrates (460-377 BCE), said, "Let your food be your medicine and medicine be your food".

Healing history

Artemisia annua is a herb in the daisy family. It has been used for over 2,000 years in Chinese medicine to treat fever, inflammation, and malaria. It contains a compound called artemisinin which has been shown to be effective in treating malaria.

Plant doctors

The bark of the white birch (*Betula alba*) contains betulinic acids, which are being studied for potential anti-cancer properties. Rattan (*Calamus*) is often used to make garden furniture but it can also be used to treat people with fractured bones. A rattan support helps the bone to regrow, because they have similar toughness and flexibility.

Rattan

White birch

Artemisia annua

Go to sleep

The mandrake (*Mandragora officinarum*) is a root that looks a bit like a person. During the Middle Ages, it was discovered that mandrakes contain a very powerful chemical called hyoscine. This drug is so strong, it could make humans fall asleep for days, so deeply that they couldn't feel a thing. The world's first anaesthetic had been discovered!

Poppy

DID YOU KNOW?
The opium poppy (Papaver somniferum) has been used as a painkilling medicine for at least 3,500 years.

Healing wounds

Sphagnum moss has been used for over a thousand years to stop wounds from bleeding. The moss is very absorbent and kills bacteria. In the present day, bandages with seaweed extract are used to help heal wounds after skin graft surgery. A skin graft is when a surgeon takes skin from one part of the body to heal a wound in another part.

Mandrake

Sphagnum moss

SPEAKING IN PLANT

Plants have been ingrained into our languages for as long as humans have been around. This is because they often form landmarks on the landscape, from big trees to bluebell woods. Today, we can still see leftovers of our ancient past in our own languages, and even some of our place names. Many place names reference something natural – look at the name 'Hollywood' for example.

Yellow rose
Friendship

Red rose
Love

Tiger lily
Wealth

Forget-me-not
Memories

Purple hyacinth
I am sorry

Flowery language

The Victorians created a language of flowers, where you could send someone a message using different combinations of flowers in a bouquet.

Flying the flag

National flags often feature plants because they are symbols of things like peace (olive branch) or strength (oak), or they might reflect the important crops and foods of that country.

Canada, maple leaf

Eritrea, olive branch

Lebanon, cedar tree

Grenada, nutmeg clove

Peru, cinchona

Mexico, cactus

National pride

Many countries have an official plant or flowers as their symbol. Indonesia couldn't decide on just one so it has three!

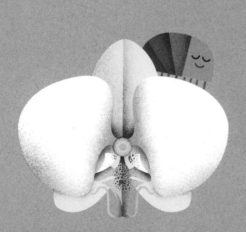
Moon orchid
(phalaenopsis amabilis)

Rafflesia
(rafflesia arnoldi)

Jasmine
(jasminum sambac)

GET GOING!

No matter how you travel, chances are that plants play a part. In the past, people only had natural materials to work with — at one time wheels and even airplane wings were made out of wood. Car tyres tend to be made out of rubber, which comes from a tree.

Buoyant bark

Cork is made from the bark of an evergreen oak tree (Quercus suber). This special bark is light, waterproof, and very durable, and protects its trees from forest fires. You've probably seen it used as the stopper in wine bottles, but it has also been used for life jackets, buildings, footwear, boats and even in the heat shields on spacecraft!

Floating villages

You've seen thatched reeds used for the roofing on houses, but did you know it can be used for vehicles as well? The Uros people of Lake Titicaca in Peru use totora reeds (Schoenoplectus californicus subspecies totora) to build boats, and they even live on floating villages made of this useful plant.

DID YOU KNOW?

An estimated 600 huge oak trees from southern England were used to build King Henry VIII's famous warship the Mary Rose. The ship was sunk in battle in 1545.

Stretch and shrink

Rubber is used to make all sorts of car parts, from tyres to the mats you put your feet on. It was first discovered by the Olmec, Maya and Aztec cultures. They used latex, the sap of the rubber tree (*Hevea brasiliensis*), to make their clothes waterproof. They even made waterproof shoes by pouring the latex onto their feet! We still have rubber on the soles of our shoes.

Fossil fuels

A fuel is something that is burned to release energy. Sometimes when plants and animals die, their bodies fall into shallow water where there is little air to help them decompose. Over hundreds of millions of years, thick layers of energy-rich materials have been crushed by the weight of the ground on top of them. This pressure has eventually led to the formation of what we call fossil fuels.

Powerful plant fuels

Oil from palm trees, algae, soybeans, and sugarcane can all be used in a car or truck's diesel engine. They can also be used for firing up a furnace or blended with petroleum to produce fuels like biodiesel. As early as 1892, German scientist and inventor Rudolf Diesel invented an engine that used groundnut or peanut oil. In the modern day, bio-gas produced by rotting plants can be used as a fuel too. With a few small adjustments, your car can be plant-fuel ready.

111

POLLUTION

If there is any **pollution** in the environment, it might be absorbed into a plant, then eaten by an animal, and then continue up the food web. This is important to know, because the chemicals that farmers sometimes use to keep their crops safe from insects, called **pesticides**, can be poisonous to people and animals. In heavily polluted land, just eleven large earthworms can contain enough pesticides to kill one robin (which can eat that many worms in about ten minutes).

Organic farming
Pollution is anything harmful that ends up in our soil, water and air. Today, there are strict laws about the safety of chemicals used in farming. Many farmers also choose to grow their crops without any chemicals at all – this is called **organic** farming.

Rachel Carson

We have the marine biologist and conservationist Rachel Carson to thank for our knowledge of chemical pollution. In 1962, she published a book called *Silent Spring*. It described how pesticides used by farmers to kill insects were entering the food chain and causing harm to animals and humans. Carson found that a single spray of pesticide called DDT not only killed insects for months afterwards, but remained in the environment long after its use. Her research was very unpopular with chemical companies, but it sparked the environmental movement.

TAKING CARE OF THE PLANET

There is no doubt that it is nicer to live and work among trees and plants. They give us the oxygen we need to breathe, make cities and homes more beautiful and can even help reduce air pollution.

Ray of hope

Some plants can suck up poisons such as nuclear radiation. In 2011, a disaster happened at the Fukushima nuclear power station in Japan that led to radiation leaking into the water and earth. 10,000 packets of sunflower seeds were sent out to be planted. The bright yellow flowers represented hope to recovering communities, but also absorbed radioactive chemicals to help make the land liveable again.

Coastal defenders

Mangroves are tropical plants with tall, branch-like roots that grow in the sea and around estuaries (places where rivers meet seas or oceans). They protect island and coastal dwellers, and literally hold the coast together so it doesn't wash away. Human-made versions of mangroves have been used to protect similar areas where mangroves can't grow (an example of biomimicry).

114

Can leaves change the weather?

In humid forests, lots of clouds above the trees are formed by **evapotranspiration** (water loss from the soil and the leaves of plants). When it rains, water flows to the trees' roots where it is absorbed, then goes up through the leaves and evaporates back into the air, forming a water cycle. If the trees are removed, the forest's water cycle is broken, and important nutrients can wash away into rivers instead of being absorbed by tree roots. If this happens too much, the whole environment may become dry like a desert or stop being able to support life altogether.

DIY: LOCAL LIVING LANDMARKS

Get to know the plants in your neighborhood through the seasons by taking part in this eco-activity. You'll learn your way around your local neighborhood and get to know its living landmarks.

You Will Need:

- ☐ A map of your local area
- ☐ A notebook and pencil
- ☐ A comfy pair of walking shoes
- ☐ A camera (optional)
- ☐ An adult to accompany you

What You Need to Do:

1. Print a map of your local area and stick it into a notebook. Zoom in as much as possible before printing – it needs to be big enough for you to add your own notes and labels.

2. Find a walking partner by asking a family member to head out with you. You'll be out for about 1 hour.

3. Choose a direction and walk for about half a mile. This would normally take about 15 minutes but you'll probably be stopping a lot to look at plants.

4. Mark the location of interesting trees and other plants you see on the street and in gardens. If you don't know what they are, take a picture or draw the leaves in your notebook.

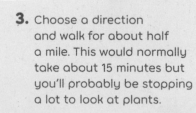

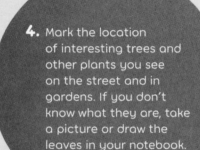

116

5. You'll see different animals on and around the plants, such as bees, flies, butterflies, and birds. Make a note of these too, to make your map more detailed.

6. Take at least one walk per season. You'll get to know what the plants look like at different times and get a good understanding of their life cycles.

7. Once you get home, check online or use an app to help identify the plants you drew or photographed.

8. Don't forget to look out for tiny plants and flowers as well as marking big trees.

THE FUTURE IS GREEN

For thousands of years, people have been looking to nature for the ultimate medicines, foods, fuels, flavours and fabrics. We will probably always do this, searching among plant life for cures and solutions to our problems. One thing we can be sure about is that your descendants and theirs too will be relying on plants for many aspects of their lives.

Home-grown

Modern farms are becoming bigger and more efficient, but many people believe that small-scale community farms and gardens are the best ways to grow food. We get about 60% of our calories from just three plants (rice, wheat and maize), yet there are thousands to choose from. Some of these alternatives include more nutritious, climate and disease tolerant crops such as algae. If you grow your own food, you can choose what to grow and eat.

Futuristic farms

There will always be wild plants, but we must make more careful use of our farms than ever before. There are already many indoor vegetable and herb farms with LED lights instead of sunlight and precisely measured doses of water and fertilizer, all controlled by computers and robots. As we become more aware of the damage that plastic has caused to the planet, will we return to natural, biodegradable materials and technologies?

Conservation

Around 21% of the world's plants are currently threatened with **extinction**, so conservation (using nature wisely in sensible, **sustainable** ways) and preservation are important. Ancient forests are being destroyed to grow palm oil that is used in making food, and this is making orangutans homeless. Tropical forests are giving way to cattle ranches for burgers, while vast fields of soybeans, sugar cane, maize and other crops are being created to make biofuels and plastics. Once thought to better for the environment (due to being 'carbon-neutral') when burnt these biofuels can be far worse in terms of how much carbon dioxide is released into the atmosphere.

Remember

If humans suddenly weren't here, plants would survive just fine, and maybe even have an easier time of things! If plants suddenly disappeared, however, humans wouldn't be able to survive for even a few minutes. We all need to learn respect for the natural world and to bring plants back into our everyday lives.

PLANT AWARDS

The oldest individual plant is the...

Bristlecone pine tree *(Pinus longaeva)*. They are found growing in the White Mountains of California — the oldest one is thought to be over 5,000 years old.

The world's slowest-growing plant is...

Lichen. Some of these colorful plants grow so slowly that it can take over 10 years for them to reach the size of a coin!

The world's widest-spreading tree is the...

Banyan tree in the Indian Botanical Gardens in Calcutta. Dating back to at least 1787, it has 1,775 support roots, and a circumference of 410 metres!

The best international traveller is the...

Sea bean *(Entada)*. It is not only one of the largest seed pods in the world (sometimes up to 2 metres across), but its shiny beans make a tremendous journey across the ocean. After floating in water for some weeks, the outer package around the sea bean decays, leaving the seed alone to float for more than a year. If it makes its way on to a warm tropical beach, thousands of kilometres away, it can still grow.

The world's tallest trees are the...

Coastal Redwoods of California, USA. One of them is 113m tall and about 1,000 years old!

The world's fastest-growing plant is...

Giant bamboo. This can grow up to a metre per day if it has everything it needs. Giant bamboo can reach heights of up to 30m and each stalk can be as wide as 20cm across – it looks more like a tree than a grass.

121

GLOSSARY

Adaptation

How a living thing becomes suited to its surroundings. This could be within its lifetime, or over many generations via evolution.

Biomimicry

Science of using ideas from nature to create new inventions and technologies.

Carnivorous

Something that only eats meat. Carnivorous, or insectivorous, plants can digest insect blood.

Chlorophyll

Green substance found in plant cells giving them the ability to use sunlight energy to make food for themselves.

Chloroplast

Structure inside a plant cell where chlorophyll is found.

Conservation

Careful, planned management of a natural resource to help protect it.

Deciduous

Trees and shrubs that lose their leaves in the autumn to protect them from freezing weather in winter.

Dwarfism

When plants are smaller than average for their species. Dwarf plants can be bred deliberately through genetic mutation.

Ecosystem

All the living things in a particular area along with all the factors that affect them (such as the weather, type of soil or each other).

Edible

Something that you can eat. This doesn't mean that it is tasty, just not poisonous.

Energy

Property of the universe that makes things move, grow and heat up.

Ethnobotany

Study of the different ways that people use plants in their lives.

Evapotranspiration

Process where moisture is transferred from the earth to the atmosphere by evaporation of water and transpiration from plants.

Evergreen

Type of tree or shrub that doesn't lose its all of its leaves at a specific time.

Evolution

Process by which organisms gradually change over generations by inheriting differences in their genetic code. This can give rise to new species.

Explosive

Something that can explode given the right trigger.

Expulse

To expel or drive out.

Extract

Substance that has been taken from something else.

Extinction

When all the individuals of a species die and there is no chance of them returning.

Food chain

Way of arranging or depicting organisms showing the order in which they are eaten.

Food web

Multiple food chains that connect to each other.

Fossil

Impression left in rock of an organism that lived a very long time ago.

Herbivorous
Something that has a diet made up only of plants.

Naturalist
An expert in or student of natural history (nature studies and wildlife).

Non-organic (farming)
Type of farming/gardening or food produced using chemicals not made from natural sources.

Nutrient
Chemical or substance that helps a living thing grow (such as minerals in the soil or vitamins in our diet).

Organic (farming)
Type of farming or food produced using only chemicals from natural sources or no chemicals at all.

Pesticide
Chemical that kills animals, plants and fungi that are unwanted. Misuse and overuse of these can lead to pollution.

Photosynthesis
Process that takes place in plant cells, where they make sugar using sunlight energy, water, carbon dioxide and nutrients.

Phototropism
Movement or growth of a plant towards a source of light.

Poisonous
Something that contains poison – a substance that can cause illness or even death if it enters the body in large enough amounts.

Pollination
Process where pollen is transferred from the anther of one flower to the stigma of another flower resulting in seed production. Self-pollination takes place if this occurs within the same flower.

Pollinator
Animal that pollinates a plant, including bats and bees.

Pollution
Presence and build-up of substances in the environment that can cause harm to living things.

Preservation
Keeping something in its original state to protect it.

Sustainable
Method of farming, harvesting or living that doesn't permanently damage or deplete the resource or the environment.

Taxonomist
Scientist who puts living things into groups and studies their origins and relationships to each other.

Temperate
Mild zone of the Earth without extremes in temperature. They are found between the tropics and the polar regions and have four seasons.

Thaw
To melt.

Transpiration
The process of water moving through a plant from the roots to the leaves, which give off water vapour.

Tropical
Climate found in regions surrounding the equator. It is usually hot and humid, with an average temperature of 18.4°C. There is a wet season and a dry season.

Tuber
Modified stem or root that stores food for the plant and can grow into a new plant.

Viscosity
How thick a liquid (or gas) is. For example, water is less viscous than honey and peanut butter is more viscous than both of them.

INDEX

A

adaptation 45, 46–7
air bladders 53
air dispersal 30
algae 12, 34
anaesthetics 107
animals
    evolution 44–5
    food chains/webs 60–1
    seed dispersal 31
anthers 22
aquatic plants 52–3
Artemisia annua 106
ash (Fraxinus) 30, 92
avocadoes 75

B

bamboo 70, 77, 89, 98, 121
Banyan trees 120–1
baobab 25
bark 77, 106, 110
bats 25
bean family (Fabaceae) 41
bean-bag boules 93–4
bees 25
Begonia 51
biodegradable
    materials 118
biofuels 111, 119
biomimicry 31, 73, 114
birch (Betula) 30, 87, 106
birds 20, 25, 45
blue dye 79
blue-green algae 34
boats and ships 110
bottle garden 26–7
bows and arrows 102
breakfast 66–9
bristlecone pine
    (Pinus longaeva) 120
burdock (Arctium) 31
butcher's broom
    (Ruscus aculeatus) 72

C

cacti 48
camouflage 103
candlenut tree
    (Aleurites moluccana) 99
carbohydrates 16
carbon dioxide 12, 49, 119
carnivores 60–1
carnivorous plants 62–3
Carson, Rachel 113
castor oil plant
    (Ricinus communis) 59
cellulose 70, 87
charcoal 70, 78
chlorophyll 16, 55
chloroplasts 16, 17, 51
citrus fruits 75
cleaning 72–3
climbing plants 50
clothing 76–9
coastal redwoods 121
coco de mer
    (Lodoicea) 29
cocoa
    (Theobroma cacao) 69
coconut
    (Cocos nucifera) 30, 77
coffee 68
conkers 32–3
conservation 51, 113,
    114–15, 119
cork 110
cornflour slime 42–3
cosmetics 80
cotton (Gossypium) 77
cricket 92
cross-pollination 24
cucumber family
    (Cucurbitaceae) 41

D

dandelion (Taraxacum
    officinale) 30
Darwin, Charles 44

deciduous trees 56–7
deforestation 115, 119
deserts 48–9, 115
diatoms 35
digestive juices 62, 63
dodder (Cuscuta) 55
drinks 68–9
durian tree (Durio) 81
dyes 78–9

E

ecosystems 51
edible plants 58, 59
egestion 31
electricity 98, 100–1
environment
    adaptation 46–7
    conservation 51, 113,
    114–15, 119
    pollution 112–13, 114
epiphytes 50
essential oils 80
ethnobotany 10
evapotranspiration 115
evergreen trees 56–7
evolution 38, 44–5
explosion 31
extinction 47, 119
extreme conditions 48–53

F

fabrics 76–7
families, plant 40–1
family tree 38–9
farming 112, 118
ferns 46, 49, 85, 103
fertilisation 24
filaments 22
flags 109
flax (Linum usitatissimum)
    76, 79
floating villages 110
flowering plants 12, 47
flowers 15, 20–5

language of 108
food
    edible plants 59
    photosynthesis 16–17
    plant-based 66–7
food chains 60–1, 67, 113
food webs 60–1
fossil fuels 111
fossils 47
Frankincense
    (Boswellia spp.) 80
fuel 111

G

germination 28–9
ginger 74
ginkgo tree
    (Ginkgo biloba) 34, 35, 81
gold 99
grass family (Poaceae) 41
grass squeaker 90–1
grassland 54
Great Dying 47

H

habitat loss 119
heliconia 25
hemp
    (Cannabis sativa) 77
herbivores 54, 60–1
herbs and spices 106
Hippocrates 106
hooks 31
horsetail (Equisetum) 34, 72
hot chocolate 69
houses 84–5
hunting 102–3

I

Indian mustard
    (Brassica juncea) 99
indigo (Indigofera) 79
ink 10, 99
insects 20, 25, 62–3

J

juices 69

jungle plants 50–1

L

Latimer, Lewis Howard 98

lavender (*Lavendula angustifolia*) 76

leaf printing 82–3

leaves 14

arrangement of 54

deciduous and evergreen 56–7

evapotranspiration 115

jungle plants 50

photosynthesis 16–17

lichen 120

light 18, 50

lotus plant (*Nelumbo nucifera*) 73

M

madder (*Rubia tinctoria*) 79

maize 59

mandrake (*Mandragora officinarum*) 107

mangroves 114

maple (*Acer*) 30, 92

medicines 77, 106–7, 118

minerals 16

mint family (*Lamiaceae*) 40

mistletoe (*Viscum album*) 55

mosses 85, 107

mummies 76

music 88–91

N

names

people 71

places 108

scientific 11

national plants/flowers 109

natural selection 44

nectar 20, 62, 63

nurse plants 49

nutrients 12, 14, 51, 62, 69, 106, 115

O

oak trees (*Quercus*) 110

oil 59, 67, 76, 77, 78, 80, 88, 99, 111, 119

opium poppy (*Papaver somniferum*) 107

organic farming 112

ovaries 23

ovules 23, 24

oxygen 12, 17, 28, 114

P

painkillers 107

paint 78

paper 10, 77, 87

parasites 55

peanuts 74

peduncles 23

perfumes 80–1

pesticides 112, 113

petals 20, 22

photosynthesis 12, 16–17, 49

phototropism 18

phytomining 99

pigments 78–9

pine (*Pinus*) 87

pistils 23

pitcher plants (*Sarracenia*) 63

plant fuels 111

plant kingdom 12, 38–9

plastics 118, 119

poisonous plants 58–9

pollen 20, 22, 23, 24–5

pollination 24–5

pollinators 20, 25

pollution 112–13, 114

potatoes (*Salanaceae*) 40, 74

R

radioactivity 114

rainfall 50, 115

rainforests 50–1

rattan (*Calamus*) 106

reeds 85, 110

reproduction 20, 24–5

respiration 12

resurrection fern (*Selaginella lepidophylla*) 49

rhubarb 79

roots 14, 28, 115

buttress/stilt 51

deep 48, 49

rose family (*Rosaceae*) 40

rubber (*Hevea brasiliensis*) 93, 110, 111

S

saponins 80

scents 80–1

sea bean (*Entada*) 120

seasons 54, 116–17

seeds 18–19, 20, 23, 24, 26, 74, 120

adaptations 46

dispersal 30–1

germination 28–9

sepals 22

shoots 28, 29

single-celled organisms 47

soapwort (*Saponaria officinalis*) 80

soya 10, 41, 119

species

evolution 44–5

plant kingdom 38–9

sphagnum moss 107

spines 48, 89

squirting cucumber (*Ecballium elaterium*) 31

stamens 22

stems 14, 29

stigma 23, 24

stinging nettles (*Urtica dioica*) 77

stomata 49

strangler figs 55

straw 84

styles 23, 24

succulents 48

sugarcane 66, 119

sugars 16, 17, 66

sundews (*Drosera*) 62

sunflowers (*Helianthus*) 54, 114

sunlight 12, 13, 16, 60

sustainability 119

sycamore (*Acer pseudoplatanus*) 30

T

taxonomists 40

tea 68

teeth, cleaning 70

temperate areas 54–5

thatched roofs 85, 110

titan arum (*Titan amorphophallus*) 81

tomatoes 75

transport 110–11

trees 12, 87

deciduous and evergreen 56–7

trophic levels 61

tubers 49

turmeric (*Curcuma longa*) 79

V

Venus flytrap (*Dionaea muscipula*) 62

volcanic ash 47

W

warfare 102–3

washing 80

water 12, 28, 48, 49

water cycle 115

water dispersal 30, 120

water plants 52–3

weeds 26–7

white birch (*Betula alba*) 106

woad (*Isatis tinctoria*) 79

Wollemi pine (*Wollemia nobilis*) 34–5

wood 84–5, 87, 88–9, 92

wounds 107

writing materials 86